Governance of Europe's City Regions

Governance of Europe's City Regions is a structured overview of current debates on cities and regions. It provides an understanding of trends at varying spatial scales and explores a range of different European experiences that consider prospects for the effective governance of city regions. Contrasts arise from diverse economic and spatial contexts, and from the complex interactions of national, regional and local politics and policy.

This book examines the trends in regional development and the responses of European scale regional and urban policy, including the changing role of the European Union with regard to regional issues. The authors' core argument develops a distinction between two types of city region, showing that city regions dominated by a single core city face different challenges from 'polycentric' regions that link groups of cities. In addition to these differences, national constitutional contexts constrain regional and local governments, and local political and cultural forces influence the effectiveness of city and regional governance. Arguments are illustrated through a range of detailed case studies, with particular emphasis on the complex interactions of regional, national and local scales in England and Germany.

Governance of Europe's City Regions introduces and summarizes a broad collection of interdisciplinary literature. It clarifies contemporary debates about regionalism and questions claims about a 'new regionalism'. The authors argue that the city region scale continues to be contested, that no single governance model will fit all experiences. They also assert that there is a need for a careful appreciation within city regions of the interactions of local institutions, their constitutional context and their economic prospects.

Tassilo Herrschel is Senior Lecturer in Economic Geography, Department of Social and Political Studies, and **Peter Newman** is Senior Lecturer in the School of the Built Environment, both at the University of Westminster.

Governance of Europe's City Regions

Planning, policy and politics

Tassilo Herrschel and
Peter Newman

London and New York

First published 2002
by Routledge
11 New Fetter Lane, London EC4P 4EE

Simultaneously published in the USA and Canada
by Routledge
29 West 35th Street, New York, NY 10001

Routledge is an imprint of the Taylor & Francis Group

Typeset in Galliard by
Florence Production Ltd, Stoodleigh, Devon
Printed and bound in Great Britain by
Antony Rowe Ltd, Chippenham, Wiltshire

British Library Cataloguing in Publication Data
A catalogue record for this book is available from the British Library

Library of Congress Cataloging in Publication Data
A catalog record for this book has been requested

ISBN 0–415–18770–2 (hbk)
ISBN 0–415–18771–0 (pbk)

Contents

1 Governance and planning of city regions

Introduction

The 'city region' has increasingly become a buzzword in debates on urban and regional development and, especially, in debates about competitiveness and processes of globalization. But what is a city region? How does it relate to the now almost ubiquitous claim of the emergence of 'new regionalism'? How is it managed and governed, and what is its role in a state structure? Such questions have become increasingly more urgent in the wake of the general resurgence, one might also say, rehabilitation, of the 'region' after its demise during the 1980s in favour of the 'locality'. What have been the reasons for this change, and how have these changes occurred under different national conditions? These include constitutional arrangements for regions and their position against local and national tiers of government, but also established attitudes to, and practices in the governance of, regions. It is these issues that this book sets out to explore and to shed some light on the intrinsically intertwined, yet also often divided, factors that affect city-regional governance. Different types of cities, their importance within 'their' regions and towards each other, all add to the complexity and the different possibilities of governing 'regions'. They can be mere territorial containers of policies defined elsewhere, or they can represent distinct identities and regional awareness. Whether this can lead to policy-making autonomy depends on the national constitutional framework and the position afforded to regions in the state hierarchy. Only by looking at individual examples of city-regional governance can difference, unique features and similarities be identified, and the relevance of the specific circumstances – external and internal – be assessed. This book has thus adopted the approach of comparative government and policy studies. Only in this way does it seem possible to study the nature and workings of city regions under different conditions and evaluate the relative importance of (a) the general constitutional provisions for 'regions', institutional practices and arrangements as 'external factors', and (b) the relationship between cities, and cities and 'their' region.

Addressing the first level of comparison, examples from across (western) Europe have been studied, embracing states with differing degrees of centralization: unitary states with continued strong emphasis on central control (e.g. England, Portugal, Denmark, Netherlands), unitary states with some signs of devolution to the regional level (e.g. France, Italy, Spain), and fully devolved federal states, such as Germany. The comparison uses a review of a number of recent case studies of regionalization in metropolitan areas as reported in the literature from across western Europe. The second part of the comparison concentrates on the two 'extreme' examples of opposite state structures, Britain and Germany. Here, two types of regions are being investigated in each country, one dominated by a large metropolis, such as London and Berlin, and one containing several smaller, competing cities. The four detailed examples thus illustrate the fundamental distinction made here between monocentric and polycentric regions. In the former case, the one major city reduces the region effectively to its functional hinterland. In the latter case, competing cities create a more complex system of territorial regionalism.

'City region', by its very terminology, means a combination of city and regional qualities both in functional and institutional-governmental terms. At first sight, monocentric city regions may suggest a greater emphasis on the local dimension through the influence of the dominant core city. Polycentric regions, by contrast, suggest more of a regional emphasis, because of the rivalry between the smaller cities across the region.

This book sets out to explore the evidence of 'new regionalism' from across the EU, so that different institutional and cultural-political circumstances can be taken into account in their impact on regionalization processes. Beginning with an overview of the main debates on, and explanations of, the emergence, nature and operation of regions, looking in particular at the respective consideration of territory and institutional governance, the following chapters explore increasingly more detailed examples of city-regional structures, governance and territoriality. These reach from a sweep across the EU, looking at examples of regionalization in different national contexts, to detailed analyses, from an international to intraregional perspective.

Chapter 2 introduces the main theoretical argumentation around city regions and regionalization, especially their characteristics and forms of governance. Much of the current debate focuses on the relationship between the fundamental societal-economic changes, seen as largely triggered by globalization, the subsequently increased interterritorial competition, and the demand for responsive and 'appropriate' governmental-institutional arrangements. This includes the mechanisms employed for designing and implementing region-focused responses, and in this the importance of relationships between different actors within or without government. Attention also focuses on the role of territory in these interrelationships both in terms of perceived common interests and policy-making capacity. The territorial boundedness of jurisdictional (or policy) areas fundamentally impacts on the scope for defining relevant and effective policies. This affects the relationship between planning and its traditionally fixed areas, and policy which is issue driven and rests on a collaboration between localities, creating 'regions' of various constellations and duration. In an ideal scenario, both would be completely congruent, but in reality this is not likely. Thus, long-term spatially based policies may achieve little more than 'hit and miss'. Scale matters here, because it will affect the likelihood of achieving good congruence between planning and policy-making region. There are differences in the ways in which inter-actor and inter-institutional relationships are formed between formal, usually hierarchical, structures and informal, non-institutionalized arrangements. This difference is also referred to as a contrast between 'hard' and 'soft' forms of regional organizational structures (Danielzyk, 1999; Priebs, 1999). The 'softer' the institutionalization, the more flexible the system becomes. It is this difference in the degree of institutionalization which is among the recurrent themes in this book, because of its central importance in the debate on 'new regionalism'. The nature of informal linkages is discussed here in the light of their contribution to the regionalization and operation of city regions.

Chapter 2 suggests a two-fold division of the main theoretical arguments: territorial (economic) and governmental-institutional respectively. The first group comprises those explanations with a distinct focus on the territorial impacts of economic change, i.e. essentially economic geography, and on the development of strategies and institutional responses within those territories. These perspectives shifted over time, with, from the mid-1980s onwards, a growing awareness of the impacts of changes in economic and industrial organization. The shifts were encouraged by the emergence of ideas of more flexible processes of production, and show a focus on how the specific qualities of places, including institutional networks, generate competitive advantage. Regions are thus seen largely as territorial containers for an, ideally, well integrated (but seemingly almost closed) system of economic and political actors, and relevant policies. The second group examines city and regional issues from a political and institutional perspective. Arguments here seek to clarify the complex relations between the three main arenas of government, nation, region and locality, and the changing relationships between public and private sectors in managing cities and regions. There is some evidence, as outlined in Chapter 2, that theoretical debates are moving from space to the wider concept of 'scale', with space merely one of the variables, and societal and political structures, the other. It is at this point that the more one-dimensional territorial debate has been broadened to also include the scale of operation of institutions. This goes beyond a merely geographic perspective and includes its internal arrangements, degree of institutional representation and visibility, linkages to within and without the government hierarchy, including vertical and horizontal interaction.

Chapter 3 examines the outcomes of changing European policies and funding regimes on cities and regions. Using the European scale, this chapter explores two main themes. First, evidence of spatial economic dynamism, in particular the changing role of city regions in the European space economy. Following the arguments discussed in the previous chapter, such changes, especially if placing greater emphasis on city regions as active economic cores, should also lead to relevant policy shifts, if policies are to remain relevant. Attention will be given to the changing nature of regions from mere centrally directed policy containers, a role traditionally associated with EU regions, to a more active role of regions as economic entities with specific indigenous development potential and a greater emphasis on cities as growth centres. The second theme is, consequently, the nature of policy responses by the EU, i.e. evidence of a greater concern with encouraging indigenous economic competitiveness of city regions, rather than merely seeking to redistribute growth. Such changes have been evident in institutional arrangements. Regions have gained a much increased presence and recognition through the Committee of the Regions, giving regions and cities a consultative role in Commission projects. The other change has been an encouragement of cross-sector partnerships, and has also had implications for national debates on regions and regionalization, with a distinct view on competitiveness in an international context.

Chapter 4 provides an overview of various examples of national responses to the raised profile of debates on regions and city-regional government. Many of the regional structures and discussions in Europe have been shaped by the impact of EU regional policy through the Structural and Social Funds. This has given national government the incentive to establish regional structures, e.g. in England, so that scope for drawing down EU funds can be improved. As a result, the picture is now more uneven in terms of the nature of regions as merely containers of centrally directed policies, or entities of self-representation and indigenous policy-making capacity. Overall, a steady move towards greater regional self-determination in (economic) policy making can be detected. While this shift was initially strongly driven by arguments about maximizing indigenous development by encouraging region-based policy making from within the regions, this has shifted somewhat to include wider issues of representation and identity, including those outside the government sphere. The result has been varied, with regions emerging as simply yet another territorial layer, while also giving much greater acknowledgement to the role of urban centres as growth areas across the EU. Government attention has shifted to city-regional relationships in the main conurbations. Traditional regional policies have been re-targeted to more urban problems, acknowledging the close interrelationship between the two. City regions, whose geographic distribution allows the identification of growth corridors across Europe, have induced new regional qualities. This, it is argued, has put pressure on existing governments, national to local, to respond to these shifts and accommodate the new city-regional functional territoriality.

The examples of regionalization in western Europe illustrate the importance of formal differences, but also point to the need to look beyond them when assessing the status of regions within the governmental hierarchies. Also, the varying objectives of regionalization become visible, reflecting the differing territorial scope of urban-regional governments and their powers available to formulate and implement regional political goals. The German city states are at one end of the scale, and the Lisbon Metropolitan Authority without independent funding, at the other. There are, however, general trends towards recognizing a greater role for regions, and this does include changes to constitutional arrangements. The main challenge, however, seems to be the boundedness of the city regions, with only a few cases where administrative boundaries and functional city regions coincide. There is considerable variation in the background to regionalization. Thus, Finland's regions emerged from EU-based policy considerations, Denmark's from competitive pressures from neighbouring Sweden, in the joint Øresund region while in France emerging regional identities (e.g. Corsica) have encouraged challenges to the centralized state structure and demanded recognized political status, including, in particular, adequate financial provisions. These challenges contrast with the case of England (although there is some similarity with Scottish and Welsh claims to independence), where no political regionalism has emerged, and the newly established structures are the outcome of a top-down policy. The importance of national identities is particularly obvious for Belgium, where cultural regional divisions threaten the state's existence. At the same time, there is 'new regionalism' emerging in the main city regions, albeit separated between Flanders and Wallonia. Thus, Brussels has, somewhat reminiscent of London and Berlin, become separated from its wider hinterland, but seeks to bridge this separation by joining less formalized city regional alliances, e.g. with Ghent and Antwerp. This form of participation in more than one region seems a specific feature of 'new regionalism', as the examples in this book demonstrate. Germany's Ruhr is another example of multilayered and multiscaled regionalization, and a polycentric region. The Randstad in the Netherlands is another example of a polycentric city-based regionalization process, albeit encouraged by the state. The latter may be one of the reasons for the quite different responses to regionalization in the Ruhr.

The forming of such city regions can be stimulated by particular events, or challenges, such as in the case of Barcelona (Olympic Games), because it reinforces local–regional identity, but it can be also facilitated by the central state to encourage nationally relevant initiatives/projects. In Portugal conditions are similar to England, where weak regional tradition encourages top-down policy implementation.

Two of the most contrasting examples of institutional and practical arrangements for regions and regionalization are provided by the unitary and federal states of England and Germany respectively. Discussed in detail in Chapter 6, they illustrate a top-down controlled form of regionalism, with all strings being held by the national government in London on the one hand, and a decentralized, inherently more regionally oriented form of federalism, on the other. With national governments being the ultimate democratic force in all countries, the main difference rests with the position of subnational government and, for city regions, the representation of local and regional interests, and its scope to reach into the region they are part of to advance local–regional interests. The balancing of local and regional interests requires particular fine tuning and may be the reason for some antagonism at this scale.

Chapter 5 sets the scene for the detailed comparison of the two very different examples of national provisions for, and recognition of, regions as part of the formal government system. In England, where a highly centralized state structure exists, with all powers resting ultimately with Parliament, from the local to the national level, there has been an interesting emphasis on informal, non-state arrangements as part of engaging the private sector. Administrative regions, essentially confronting local and national democracy, and the plethora of quangos dealing with urban-regional matters may be seen as a substitute for more genuinely devolved democratic control. Regions have not featured in England as part of the government hierarchy at any time in its history, and only the recent devolution to Wales and Scotland shows a recognition of their national identities and a step back from the centralization of the United Kingdom. At the practical level, regional considerations have some recognition in the shape of the counties as upper tier of local government, but this is very limited. Otherwise, regions have not been more than containers of centrally defined and implemented policies, whether based on Keynesian ideas of the 1960s, or the more recent paradigm of globalization as all-embracing parameter of societal and state activity, enforcing territorial competitiveness. Except for London, there is no democratic representation at the regional level, and no regional government. The emphasis is on private sector style, marketing oriented Regional Development Agencies as the 'face' of the regions, operating primarily under central government control and having only very limited region-based legitimacy. Their backbone is central government, through its regional offices not the regions.

This situation contrasts sharply with that in Germany with its federal constitution, and devolution of most home affairs to the regional states, the *Länder*. They act statutorily as central governments. This empowered regional status is accompanied by a complex, hierarchically and highly formally organized state structure and distribution of responsibilities. Also, there is a dual hierarchy of policy responsibilities and territorial planning. The latter enjoys a central role in public administration and government. Territorial planning for regions operates at two levels, the *Länder*, and the planning regions at sub-*Land* level. They are administratively created 'containers' of *Land* originating policies, and subdivide the larger democratically controlled *Länder*. This, and a detailed provision and regulation of policy making and responsibilities, adds to a lack of complexity and claims of lacking flexibility and responsiveness. Competing responsibilities add further to a sense of separateness. Reflecting differing values and traditions, each *Land* placed its own emphasis on more centrally or locally arranged planning regions, but there is a growing recognition that bottom-up regionalization allows better response to economic geography and greater responsiveness per se. This experience is beginning to feed back into the strictly formalized state-administrative structure. There seems an interesting parallel shift towards more flexible forms of regionalization, away from territorially fixed, institutionalized arrangements to more informal, non-institutional and inherently more dynamic approaches. While in England such a move has resulted from continued central control and little practical devolution, in Germany, the driving force has been too much, often obstructive, and rigid institutionalism.

Chapter 6 investigates the evidence of the impact of these differences on regionalization processes in city regions in more detail. This includes, in particular, evidence of a positive relationship between the degree of centralism and absence of regional self-determination in policy making. Does such a relationship entail a more coherent 'streamlined' relationship between spatial tiers of government at the expense of individuality? Is decentralized government 'more messy'? More detailed evidence suggests a more varied picture. Approaches to regions and, especially, regional policy, depend on particular local circumstances, including personalities and established practices, established 'divisions of responsibility and power' both institutionally and territorially, and democratic control and credibility of institutions and actors. Also important are established attitudes to the desirability, and thus importance, of regionalization of state power and governance both from a local and national perspective (as well as from the region). Thus, in Germany, regions are an inherent statutory and cultural-historic part of state consciousness, while challenged by the federal and strong local level of government. Also, there is a strong tradition of hierarchical spatial planning with a clear allocation of powers and responsibility to different tiers of government, providing the 'backbone' to the inter-operation between government scales. Nevertheless, the system is complex, and competing hierarchies of policy making and territorial planning may emerge, each claiming greater constitutional legitimacy and importance.

In England, all forms of governance are ultimately controlled by the central state, issuing guidelines and finance for policies and thus establishing how much regionalism there is at any one time and how it is to be applied. This may include establishing new players such as the Government Offices in 1994 or Regional Development Agencies in 1999 and engagement with

EU institutions. Nevertheless, regions and city regions in both countries are seeking direct representation in Brussels and inter-regional collaboration in lobbying may transgress national divisions. In both countries there have been interesting similarities in discussions on regionalization especially the relationship between cities and their wider hinterland region. There is also the issue of boundedness, i.e. the appropriateness of territorial delimitation of regions for their tasks, and the growing number of non-institutionalized policy territories such as employment offices, chambers of commerce and single purpose local government associations (lobbying groups) add to the complexity and potential obstructionism (institutional inertia) which exist already through a plethora of layers of (non-congruent) government territories. In Germany, for instance, there are planning regions within each *Land*, themselves also referred to as 'regions' from an international (EU) perspective. The planning regions are not part of the government hierarchy and may be established top-down by a *Land* or bottom-up through local cooperation.

Against the background of distinct (and considerable) differences in provisions for regionalization in England and Germany, there are strong indications that regionalization in both countries can operate through different (vertically and/or horizontally parallel) avenues of region building: directed and controlled 'from above', and shaped and utilized 'from below', as vertical interaction, while operating horizontally through the links between various actors be they competitive or cooperative in their attitude.

Overall, irrespective of national contexts, there seems to be a growing presence of informal (non-institutionalized) network-based forms of regionalization, both at local authority level (inter-local collaboration) and institutional (personal) level between different actors as part of governance. Both promise a more distinct and publicly supported regional identity than if imposed top-down.

Chapters 6 and 7 explore in detail regionalization processes by comparing city regions with different internal structures, and external arrangements and provisions. Differences in internal structures revolve around the distinction between monocentric and polycentric city regions. In the former case, one major urban centre structurally dominates the wider region which is little more than its hinterland. In the absence of equal competitors for influence, the core city's interests push for recognition throughout the region, but often there is considerable resistance by the region against such domination and perceived imbalance of interest. Institutionalized boundaries between policy territories may be used as lines of defence of interests and challenge rather than encouragement to collaboration. London and Berlin are two such examples. They both represent the top positions in the respective urban hierarchies and dominate functionally a much wider region. In London's case, new territorial divisions have been established by central government's decision to establish three new regions cutting across the London region, through established subregional territories of informal collaboration, and separating London from its hinterland. It is this process which brought the London region closer to that of Berlin. There, Berlin and surrounding Brandenburg are separated by 'high' *Land* boundaries, separating the two policy arenas. But in Berlin there are ongoing political debates about the unsustainability of this division, while in London a similar situation has just been created. But while in Berlin's case there is at least one administration for the whole of the region surrounding the city, albeit subdivided into 'planning regions' with no administrative powers, in London's case the three regions are actively encouraged to compete and thus implicitly confront each other as part of the whole rationale of regionalization as a means of increased economic locational competitiveness. No provisions have been made for bringing the competitors together. Regionalization in London's case has thus increased the amount of institutionalized separateness (boundaries) across the region, while also raising the divisions through encouraged competitiveness. The result ultimately will be non-cooperation in order to maximize one's own region's interests, not least, as in the case of the RDAs, to be seen to be successful in the eyes of the supervising government. Financial reasons are the main reason for non-cooperation in Berlin, where each 'side' is watching the other, jealously looking for possible 'drains' on its resources through services used by the other region's population, or opportunities through business investment. Both city regions thus illustrate the potentially divisive nature of institutionalized territoriality. Thus it is not merely the very existence of these boundaries, but the political and financial values attached to them, that define their effect and intransigence.

The issue of competitiveness and envy between, but also within, regions is illustrated in Chapter 7 for the two polycentric examples discussed, Yorkshire and the Humber in England, and Saxony-Anhalt in (eastern) Germany. Here, the English region and eastern German *Land* respectively, encompass four and three competing large cities each, challenging somewhat the notion of a coherent regional entity. In the case of Yorkshire and the Humber, this rivalry, can go as far as mutual accusations of damaging the other's interest, as in the case of Leeds. The region, for instance, views relatively successful Leeds with envy and accuses it of 'sucking the region dry' of investors. These divisions also operate at the next smaller scale, where the subregion of South Yorkshire accuses its main city, Sheffield, of hoovering up all EU grants. Such mutual recriminations do little for a sense of regional common purpose. Collaboration only seems to be possible when incentivized, such as in the form of EU monies, or coerced through government policies and directives. Yet, while in Saxony-Anhalt there is now a process of locally inspired re-regionalization to reflect internal divisions and economic geography, encouraged by the *Land* government, no such attempts exist in Yorkshire and the Humber, where it takes funding opportunities to entice communication and collaboration. The rivalry between the four main cities does little to dispel divisions, and questions the RDA's scope for successful operation.

The changing role of regions established by the Labour government appears to have done little to facilitate a common regional purpose agreed by all main players other than maximizing return from financial incentives (grants). The continued underlying centralism – RDAs are answerable to central government – and competing involvement by different government departments promises little in terms of regional coherence. It is the main cities and the Government institutions (and RDA) that seem to set the agenda, even if negative, leaving the other parts of the region much less visible (and represented). It is here where the German example offers somewhat more scope for regional representation through informal local groupings, also outside the main cities, and their recognition as part of the region's governance. The two types of region clearly illustrate the multifaceted nature of regionalization and that certainly 'one size does not fit all'. In polycentric regions the scope (and incentive) for cooperation between otherwise competing cities needs to be accepted by them and translated into agreed policies, if benefits are to be achieved from regionalization. In monocentric regions the playing field seems to be clearer, if less level for the regions themselves. The dominance of one major city acts as somewhat of a guidance of interests, as regional welfare largely depends on it (even if not admitted). So, different mechanisms and objectives may be required to achieve the relatively best outcome of regionalization for the cities and regions involved.

Chapter 8 summarizes the main findings of the case studies at the different spatial scales. There are some interesting similarities, but also distinct differences, in response to varying traditions, established practices and institutional arrangements. Nevertheless, there is a strong indication that European regional governance is likely to develop further in multilevel and asymmetrical directions. Financial provisions as the ultimate determinant of true regional policy-making capacity repeatedly emerge as a crucial factor in the operationalization of regions. In England, establishing the new regions has been largely symbolic, as real power, especially finance, is strictly retained by central government. Effectively, the buck of responsibility and taking flak for perceived failures has been passed down, but not the means to really do something about it. Requiring regions to compete for inward investment is likely to increase non-cooperation between regions. The danger of financial considerations leading to the building of new walls is illustrated by the Berlin example. Concern for maintaining financial income, and minimizing expenditure, has fundamentally jeopardized potential for cooperation. It remains to be seen to what extent the key feature of new regionalism, the inclusion of new actors as part of a shift of governance, can help to overcome the institutionally-based divisions and competitiveness. The importance of financial incentives, and the political pressure to be seen to succeed, have resulted in pragmatic regional collaborations. These may not outlast the life of the incentives, but they essentially reflect the nature of new regionalism as a flexible, purpose-driven and inherently unstable arrangement. The cases discussed in this book point to two important lessons that can be learned: the need for institutional cooperation across politically divisive borders, and the importance of clear, leading images underpinning effective regional policies and their acceptance among other players in the government hierarchy.

2 Theoretical explanations of city regions

Territory, institutions, networks

The pace of change in urban and regional governance has been matched by rapidly expanding bodies of theoretical work that seek to explain the forces of change and their direction. In this chapter we explore a range of debates about urban and regional change and accounts of the changing nature of city and regional governance in Europe. Theoretical perspectives on city regions draw on a variety of disciplines. We select a number of important issues from geographical and political science perspectives. These disciplinary boundaries are not hard and fast and common issues come to the fore of debate across the social sciences. There is interest in the social and economic impacts on cities and regions of these changes, and in the implications for the institutions of governance that attempt to steer paths through economic uncertainty. Perhaps it is not surprising that perspectives on these changes are not fixed and there is considerable debate about the usefulness of differing theoretical frameworks (see also Aschauer, 2000).

The arguments can broadly be divided into two main groups. The first comprises those with a focus on the territorial impacts of economic change, on development strategies and institutional responses. These perspectives shift over time, with, from the mid-1980s onwards, a growing awareness of the impacts of changes in economic and industrial organization, i.e. the emergence of ideas of more flexible processes of production (Admin and Robins, 1990), and focus on how the specific qualities of places, including institutional networks, generate competitive advantage. One of the important theoretical debates behind much of the debate in this group is the shift from Fordism to post-Fordism in regulation theory (see e.g. Amin, 1994). Theorists argue that new forms of industrial organization in post-industrial societies are paralleled by changing forms of state regulation and intervention, and will also shape urban development (Lipietz, 1992).

The second group of arguments examines city and regional issues from a political and institutional perspective. Arguments here seek to clarify the complex relations between nation, region and locality and the changing relationships between public and private sectors in managing cities and regions. Core theoretical debates focus on a transition from government

Economy and Territory	Territory and Governance
Globalization economic competition and policy choices	**Rescaling the state** the interaction of economies, institutions, identities
Untraded interdependencies firms and soft institutional infrastructure	**Multilevel governance** tensions between national and regional identities
Industrial districts reflexive and learning regions	**Growth politics** 'innovative milieu', urban boosterism
Post-Fordism and regulation theory flexibility, specificity, individuality	**Governance and networks** uneven institutional development
The New Regionalism the normative city region – learning from success?	

Figure 2.1 Urban and regional theory – mapping the debates.

(concentrating on formal institutions) to governance (more flexible, networked arrangements involving private as well as public actors) and on the 'rescaling' of states that can be seen in both a weakening of the traditional roles of nation states and increasing importance of regional and local scales.

The challenge for both groups of arguments is the inherent ambivalence in the nature of regions, being situated between the local and national level. Regions are recognized as economic hubs in a global economy but there is a fundamental lack of clarity about the functions of these hubs and their relationship to local, national and global scales. Blotevogel (2000: 496) refers to the region as a 'multi-dimensional semantic field' with 'fuzzy edges' and a 'multi-dimensional meaning' as established by the different users and analysts of 'region'. He sees the region described by its spatial and scalar reference, and its (functional) purpose. The indeterminate region is also subject to competing claims of influence from the 'top-down' policies of central governments and the 'bottom-up' claims of localities. The region has been understood as a fixed container of nationally defined initiatives of spatial growth management and distribution. More recently, regions have been represented as dynamic spaces locating themselves in a global economic system with leading cities making the case for less centralized economic management, and where institutional and functional interrelationships become decisive factors (see also Blotevogel, 2000; Cox, 1997). We need to understand how regions have become active players in the process of transformation.

The regional and urban scales of analysis have followed different pathways. Europe as a space of regions has been a long-standing focus for geographers. Over the past few decades the main questions have been about the convergence or not of regional fortunes and about the dynamics of successful and of failing regions. The beginnings of a different perspective can be seen in the late 1980s when the EU undertook a series of studies of European cities. This work located cities in regional contexts but also sought to analyse specifically urban factors that made some cities more successful than others (Parkinson *et al.*, 1992). Urban analysis drew on different academic traditions to regional studies and in particular sought explanation of relationships between government and other actors as part of 'urban governance' and the attempt, for example, to build 'growth coalitions' to maximize local development potential and opportunities, including 'boost-erist' urban policy. Understanding changing European space and governance includes both an urban and a regional dimension. We focus on changing economic space in globalized markets and on the contested political space of European city and regional governance. Cities are now seen as the motors of regional economies (Hall, 1998) driven by new forms of governance. What links these perspectives in both academic and policy discourse is the context of globalization.

Territory-focused concepts of economic change and institutional responses for the regions

Globalization and the place of cities and regions

Urban and regional analysis was profoundly influenced in the 1990s by the dominant discourse of globalization. Two strong lines of influence are those which link a globalized economy to regional and to city-regional economies and those which link the goal of economic competitiveness to necessary reform of city and regional governance. A starting point for exploring these debates is with studies of the fundamental impact of a globalized economy on the system of cities and regions.

The world systems literature takes cities out of their former national context and regional context and locates them in new networks. Some cities have exceptional roles and Beaverstock *et al.* (1999) identify numerous (50+) 'world cities'. The function of London is so exceptional that it can be argued to be not a European city but related much more to New York than to the UK or Europe (Taylor and Hoyler, 2000). While it is possible to identify new roles in managing global markets or flows of labour, or cities having an internationally recognized cultural standing, locating cities in new international hierarchies can overplay their functional roles at the expense of under-standing the historical and cultural factors which go to shape them (Abu Lughod, 1999). The world city idea also underplays the distinctiveness of groups of cities. Le Galès and Lequesne (1998), for example, highlight the distinctive character of European cities. Interpreting globalization as a com-petitive interurban or interregional race in which global functions define city-regional roles needs careful handling. Cities and regions respond in different ways and economic competition is only one explanation of the emerging relationships between places and their global context. Clarke and Gaile (1997) talk about the range of 'causal stories' that are used to link the local and global. Economic competition is one of them, but there are other, environmental for example, accounts of this relationship. The extent to which regions are free to negotiate relationships between the local and the global is a big issue and one that we will return to throughout the book.

The dominant account of globalization as an economic force is itself contested. The economics of globalization are contested in terms of international patterns of trade, flows of capital and the impacts of economic competition on urban and regional systems. Hirst and Thompson (1999) argue that the increasing internationalization of trade and financial flows is largely concentrated in the developed world and in distinct trading blocs rather than being truly global in reach. It is also clear that we cannot distinguish the economic facts of globalization wholly from the international policies which are supported by nation states. Indeed it is the nation states themselves that agree the treaties to free up trade and encourage a form of economic globalization. Thus a core question arises about how much global economic change is pushing cities, regions and nation states in similar directions or how much scope there is for public policy to intervene and change or divert the course of international pressures. John Friedmann, who has been contributing to the world city debate for over twenty years, firmly believes that public policy choices remain open to decision makers (Friedmann, 1999). We will return to this fundamental issue.

Globalization has implications for city economies and a changing hierarchy of cities. An alternative perspective on global trends locates regions at the heart of structural economic change. Scott (1998) constructs a picture of a global economy made up of 'regional motors' surrounded by 'prosperous hinterlands' (Tokyo-Osaka, Southern California, the Milan-London 'banana') and extensive frontiers to be exploited. The contemporary geography of global capitalism can thus be shaped into regions with dynamic economic cores.

An important part of this argument is the proposition that cities and regions have become disconnected from their national contexts (Scott, 1998; Barnes and Ledebur, 1997). National economic space has become less significant in a regionalized view of economies tied into new global networks of relationships. Thus the significant impact of global economic change is at the regional scale where we are asked now to imagine core cities as economic drivers linked in regional clusters that dominate global networks.

There is some backing for this perspective from evidence of greater volatility of economic performance of urban areas (Lever, 1997). Lever presents suggestive information of the gap between national economic performance and the fate of urban economies. From this evidence the economic performance of cities appears less related to national economic fortunes. Cities do well or do badly in terms of their own response to economic competition. Also, urban economies rise and fall in relationship to others over very short periods. Given this context of increasing volatility it is not surprising that city governments develop competitive strategies that may, in turn, increase differences with national economic performance. Numerous league tables of economic performance, produced either by economic geographers or by business journals, have become important reference points for policy makers and marketeers. The differences between the tables are of less significance than the proliferation of this type of measure of city and regional well being. Since the 1980s the competitive ambitions of cities in a global marketplace have been transparent. Marketing and promotional activities, the pervasive slogans of competitive cities and competition for prestige events all demonstrate an underlying concern with competitive position on an international scale. Not all promotional activities are successful and most make little sense from the perspective of national or European competitiveness. Cheshire and Gordon (1996) argue that city marketing within the single European market is a zero sum game; it can be argued that the sum of the economic performance of volatile urban economies does not add up to increased growth in a national or in the European economy.

But, as we have already noted, economic competitiveness is not the only link between the global, and city and regional levels. Environmental issues are seen as global both in their impacts and solutions. The policy options for cities and regions have been marked by environmental globalization that has, 'shifted the balance of power in the formation of the political agenda' (Held *et al.*, 1999: 410). In Europe, environmental standards and policy frameworks have a long history and have substantial impacts on public policy choices. But environmentalism remains contentious and debate about environmental issues is most often expressed as a conflict between neo-liberal economics and environmental sustainability. Within regions, environmental problems are not evenly distributed. Those with less choice in labour and housing markets are likely to be additionally disadvantaged by differences in environmental quality. These distributional issues have brought forth debates about environmental justice (Keil, 1998; Harvey, 2000) and the need to understand regional linkages that often underpin environmental issues. Environmental issues often have a larger than local scope. Transportation, waste disposal, and planning for growth obviously have regional dimensions. In the US, debate about 'smart growth', development that minimizes environmental impacts, is most often tied in with ideas of regional governance reform (see Pastor *et al.*, 2000). Keil (1998) explores the ways in which environmental campaigns take on a regional dimension. Thus the environmental agenda has impacts on the institutions of regional governance. Not only has environmentalism spurred international treaties and the enforcement of national standards and policies but environmental issues can also demand solutions at the level of regional government.

Both the economic and environmental dimensions of globalization debates point to important institutional changes. International organizations such as the World Bank, OECD and the UN, through its Habitat initiative, are all concerned with quality of urban and regional management as cities and regions adjust to global economic change and seek solutions to environmental problems. The 'structural adjustment' of public sector management and policy demanded by the World Bank includes less bureaucratized government and more involvement of business in decision making (Harris, 1997). In 1999 the OECD produced a set of principles of metropolitan governance (OECD, 1999). They argued that as metropolitan areas play an increasingly important role in the global economy, the way in which they are governed has become crucial to their ability to grasp economic opportunities and resolve questions of social cohesion and environmental sustainability (see also Gödecke-Stellwann *et al.*, 2000; Healey *et al.*, 2000). From its international perspective, the OECD identified widespread problems of lack of accountability and transparency of decision making. The challenges of economic change therefore have institutional consequences. Some strands of debate (for example, Scott, 1998) suggest an inevitability about the emergence of appropriate new institutions that are functionally required to meet new regional economic or environmental challenges. However, other views suggest that city and regional governments are not being forced in particular directions but that national and city and regional governments face choices (Freidman, 1999). To see regional governance as a functional imperative in a globalized world is clearly overstating the argument. How far particular cities and regions take on international agendas is a matter of choice, even if choices are constrained. The variety of responses to economic and environmental pressures undermines the idea of functionally necessary government. Keil (1998), for example, argues that new regional politics in Los Angeles are a specific response to global city pressures. Soja (2000) argues that while new forms of regional cooperation can be seen as a response to the current phase of urbanization, regional politics can be either progressive, as in the case of public transport lobbies in LA, or regressive, as in the moves by some favoured localities to secede from regional government. What these examples, albeit at some distance from Europe, offer is a clear view of the range of policy and institutional consequences of economic and environmental pressures. At international, national and city-regional scales political leaders and communities struggle to come up with effective institutional and policy responses to perceived global issues. Institutions do not adapt quickly, and sub-optimal decisions are always possible. Thus while global economic and environmental pressures are keenly felt there is no inevitability about institutional and policy responses. We need to look at a range of other types of explanation about how cities and regions actually change their policy directions and how the institutions of governance develop.

We now shift our focus from global scale to localities. This is a reverse of the coin, looking at cities and regions not from the perspective of broad global forces but from literature concerned with explanations of local difference.

Territory and economy – regions as economic spaces

Perhaps the most significant shift in economic understanding of regions comes with the insights of Storper (1997) and others into the 'soft' characteristics of successful regions and the 'untraded interdependencies' which embed economic activities. These non-traded characteristics include labour markets, public institutions, customs and values (1997: 19). Economic geographers have used this and similar ideas (for example, 'institutional thickness', Amin and Thrift, 1994; Cooke and Morgan, 1994) to identify the qualities of successful regions. For Storper, economies, organizations and territories are intimately related. The competitiveness of firms depends in part on the qualities of their location and, in the contemporary economy, with its emphasis on innovation and learning, the 'region' becomes part of the 'supply architecture' for success (Storper, 1997: 22). Regionally specific assets provide the conditions in which companies and economic sectors thrive.

For some years economic geographers have researched not just the economic advantages of regions but the institutional factors that contribute to success. For example, the corporatist network relationships effectively define the 'industrial district' identified as the 'Third Italy'. Regions defined in this way become the smallest unit of economic interdependencies and the idea of industrial districts made up of networks of small firms is a core element of claims for a new regionalization of the European economy. The cases of Emilia-Romagna in north-east Italy and Baden-Württemberg in southwest Germany have been discussed repeatedly as exemplars of new flexible economies based on indigenous, specific regional economic potential and also effective institutional capacities (see Heidenreich, 1996; Koch and Fuchs, 1999).

> Industrial districts [were] characterised by cooperative industrial networks and supportive institutional environment [and] were regarded as guarantors of high employment and income levels, economic growth, increasing export rates and high tax revenues.
>
> (Heidenreich, 1996: 401)

Economic indicators appear to support the geographers' analysis. In the early 1990s when these theoretical insights were being debated, these two regions were among the top 'performing' regions of the EU, with a GDP about one third above the EU average (Heidenreich, 1996). Since then, however, the success story appears to have run out of steam with considerable increases in unemployment and indications of structural economic difficulties, reflecting the changing economic parameters set in train by globalization. The successful industrial district rose to the challenge of overcoming outdated modes of production, but growing international competition has changed the economic context. Consequently, policy responses need to be found to restore, and/or maintain, regional competitiveness.

In Emilia-Romagna, the regional economy is a patchwork of local indus-trial districts, often dominated by one industry. Baden-Württemberg is characterized by larger firms and boasts a more genuinely regionally organ-ized economy, driven by a strong *Land* government adopting a distinctly corporatist approach through established and government-institutional structures and linkages (Cooke and Morgan, 1994). This 'local versus regional integration . . . is connected to a completely different institutional setting' (Heidenreich, 1996: 407), contrasting informal, 'friendly' inter-company relationships, and relatively limited class polarization (Scott, 1988). The involvement of Chambers of Commerce, state agencies and local government seems to provide the necessary stability and degree of conti-nuity as a basis of strategic policy decisions. Institutionalization plays a much greater role in Baden-Württemberg, reflecting the traditionally strong cor-poratist relationships between government and economic actors at all levels of government, 'Therefore, the crucial problems in the German region are linked to the reform of established institutions – and not the creation of new ones' (Heidenreich, 1996: 408). Regionalization is here much more based on institutional structures, providing a corset-like framework for economic activity, whereas in Italy, regions are defined by inherently flex-ible and varying economic linkages within and between localities as modules within the region. The links between organization, economy and territory are evidently complex and follow specific routes in different parts of the European economy.

The literature on industrial districts brings to the fore questions of the complex patterns of interrelationship and local choices. However, a note of caution is needed here as these types of relationship and forms of industrial district are by no means widespread. The quoted cases may well be excep-tional (Hudson and Williams, 1999) and such clusters of activity may well fail to continue to generate the institutional innovation needed for continued success.

At the heart of Storper's analysis of regional development are the ideas of learning and innovation. The current phase of global economic develop-ment seems to demand reflexive actors in both public and private sectors who can ensure supportive institutional environments. The idea of the 'learning region' (see Florida, 1995) draws on the idea of untraded rela-tionships and promotes a style of regional economic development based on the systematic use of complementarities to exploit a bottom-up develop-ment of enterprises and institutions. The basis of regional competitiveness, according to Florida, includes continuous creation of new knowledge-based production and the education, training and networking between firms and public agencies that will sustain competitive advantage. The focus is on small to medium-sized enterprises (the backbone of economic life in most

European regions). A key feature of the learning region is organization through networks that bypass bureaucratic controls. The challenge for successful regional development lies in the integration of training and enterprises. The idea of the learning region continues to promote the core beliefs of the economic geographers, that business and government need to interact in new purposive and informal ways.

This analytical linkage between firms and institutional networks chimes with more generalized perspectives of the future of the European economy. Veltz (2000), for example, argues that competitiveness is increasingly 'environmental', meaning that cultural and geographical factors play an important part in creating competitive edge. The underlying network model of economic organization points to the economic advantages of the vast labour markets of large agglomerations where the right infrastructure, including the soft infrastructure of urban governance, can make a difference.

Successful regions demand flexible policy responses and also flexibility in governing institutions themselves. The idea of networks of relationships between firms and between governance institutions and across sectors lies at the heart of work on flexible regions and industrial districts. We return later to the importance of network theories to understanding the changing governance of cities and regions. An important aspect of this debate is the contribution of informal relationships to the dynamism of regions. The formal institutions of government are important, but other relationships also seem to play significant roles. This includes the importance of 'milieu', or business environment, and the quality of its inherent creativity and responsiveness (Aschauer, 2000).

These perspectives on industrial districts, learning regions and soft regional infrastructure, look up from the often dense and informal patterns of relationships between enterprises and government. They start from a concern with the adjustment of regional economies to the radical changes evident in Europe's traditional industrial areas since the 1970s. More recently, an increasing regionalization of the European economy has been deduced from new perspectives of global economic change. These changes in industrial organization and supporting infrastructure can be subsumed within the broad sweep of regulation theory (see Boyer, 1990). The core of the argument is that the state responds to economic change by developing new forms and techniques of economic management. When applied to the subnational scale (Painter, 1991) changing regulation can be seen in the shift from bureaucratic organization to the more flexible structures that support flexible economies. While these debates echo the concerns we have discussed in this section, the regulation theory debate is not without conflict. Painter (1995) points out many of the problems arising from applying regulation theory at subnational scale. Indeed, regulation theory takes us into a discussion of whether or not a fundamental change from Fordism to post-Fordism has in fact taken place, but nevertheless the accounts of changing relations between territory, economy and institutions that we have examined seem to offer numerous and useful insights into change without getting embroiled in such theoretical argument.

Our next section examines recent work which locates this economic regionalization in substantial changes in patterns of governance.

City and regional institutions and governance

Regions as part of the 'rescaled state'

Academic debate on the appropriate territorial 'scale' to manage the economy in a global system and debate about appropriate scale of intervention by the state has gained in momentum over the last few years (MacLeod, 1999, 2000; Jessop, 1994, 1997; Jones and MacLeod, 1999; Jones, 1998; Scott, 1998; Swyngedouw, 1997, 2000). This debate points to the mutation of established state organizations and territorial structures to include new scales of government (either in addition to, or in substitution of, existing ones) which are deemed more appropriate for responding to changing societal-economic arrangements. The emphasis is thus on changing relations between social and economic territoriality and the adjustment by the state to these changes.

The academic roots of these debates are in discussions of globalization and its continued impact on national and subnational territorial structures, identities and governments. Change is characterized by increasingly firmer territorial global–local interrelatedness, and this has encouraged a new set of concepts such as 'glocalization' (Swyngedouw, 1997) and 'hollowing out' of the nation state (Jessop, 1994). Both concepts are underpinned by the notion of shifting territoriality in governance, in response to outside (primarily economic) stimuli and a growing emphasis on the subnational, especially regional, level.

Building on the work of the French urban theorist, Lefebvre, Brenner introduces the 'principle of superimposition and interpenetration of social spaces' (Brenner, 2000: 370) and suggests local (urban) transformation processes as, 'expressions of a multiscalar reterritorialization of intertwined geographical scales' (Brenner, 2000: 370). Brenner paints a picture of radical transformation. These changes in space pose challenges to the operational arrangements of the nation state and associated territorial government. It is the regional tier that seems to be the main arena of rescaling as government regulation moves upwards or downwards. The new interest in regions can be seen as 'the confluence of these processes of functional and institutional restructuring with political mobilization in the regions themselves' (Keating, 1997: 388).

The debate about scale thus focuses on the dynamism of the scalar arrangement of the state, including changing boundaries of governing institutions and changing responsibility between established and newly emerging government levels as part of a 'scalar flux in which interscalar hierarchies and relations are continually reshuffled in response to a wide range of strategic priorities, conflicts and contradictions' (Brenner, 2000: 373). Economic change impacts on space, creating new institutional challenges.

The intermediate scale of the region comes to the fore as economic change focuses on urban agglomerations where there is increasingly more uncertainty about the scale of large urban areas, or city regions, where local and regional scales meet.

The argument about territoriality and appropriate scale also draws on debate about the changing nature of capital accumulation and the 'profound geographical reorganization of capital' (Harvey, 1995: 5) and a revisiting of discussions on the making of economic spaces in general, and that of regions in particular. At the heart of this discussion is the contrast between the notion of regions as territorial containers of policies with a traditional emphasis on fixed and clear territorial boundaries, and more dynamic, social spaces.

These debates suggest some fundamental rethinking of urban and regional theory. Brenner (2000: 367) identifies two main roots of the 'recent scalar reformulations of the urban questions' (367): (1) the claimed increasing reorganization of established, historically shaped, often entrenched, relations between urban and supra-urban (primarily regional) scales of political-institutional representation (increasingly network shaped), and (2) the beginnings of a 'reconceptualization of the nature of geographical scale itself as an arena, hierarchy and product of capitalist social relations'. This includes the understanding that 'each geographical scale operates simultaneously as a presupposition, a medium and an outcome of social relations' (367). Effectively, therefore, this understanding presupposes a production of space (Harvey, 1996, 2000) through social processes, i.e. the delimitation and operation of territoriality in response to societal pressures and processes. Such production will inevitably include varying scales of state and, eventually, government, each seeking to exercise their influence and interact with, and respond appropriately to, changing sociospatial structures from the national to local dimension. Regions may be argued to be a favoured scale. They are situated in the scalar 'middle' between 'local' and 'national', and an inherently unclear definition of their own scale that makes them appear predisposed to accommodating newly emerging, changing scales of governance. This theoretical work suggests that the concern with the regional scale is not a temporary fashion but tied into more profound social and economic transformations.

The concept of rescaling government explicitly includes questions about identity and representation at various economic and governmental scales. MacLeod (1999) points to the representational aspect of scale, i.e. that each scale of governance has its associated social activities and related political-institutional representation, aimed at its reproduction at that same scale (Jones, 1998). At base, the argument leads from the spatial division of labour to a spatial division of differently scaled regulation areas both horizontally and vertically organized. The map of Europe could therefore be seen as a patchwork of territories with changing boundaries and a sequence of vertical tiers of differently scaled government territories.

Vertical layering of various scale territories does not, however, mean simply a 'territorial scaffolding' (Brenner, 2002: 15) as a vehicle for the articulation of processes and developments, and responses to them, but it also shapes institutional capacity and the ability of policy to have an influence. In effect, new, differently scaled territories of state and society can be added on to old spaces and their respective institutional practices. The creation of ever more quango-based territories (regions) of responsibility, such as in England, is one example of simply adding territorial layers; the superimposing of historically derived territories of different scales in Germany, another. Against this background, Peck (1998) emphasizes the importance of the interaction between 'the unfolding layer of regulatory processes/apparatuses and the inherited institutional landscape' (29) for developing particular and changing geographies of state power. This includes the changing roles or regional governance within the spatial hierarchy of government. The outcome may be a growing 'thickness' (Amin and Thrift, 1994) of institutional and territorial constructs, superimposed, often incongruently, in the wake of attempted responses to the perception of changing tasks. We may assume that outcomes will not always be effective in managing regional economies.

Out of these, at times, dense arguments about rescaling comes a series of useful insights into changing patterns of governance. Responses to changing economic geographies and sociospatial characteristics appear to favour a shift away from centralized structures towards multicentric power structures (after Brenner, 2002: 26). An outcome of such scalar expansion (both horizontally and vertically) may be a loss in coherence and efficacy of policies and governance, with a growing danger of competition between a multitude of institutions.

Not surprisingly, with such complexity and dynamism, Swyngedouw (1996: 1500) observes that despite a lively debate about state rescaling, 'the actual mechanisms through which this takes place remain vague and undertheorised'. The examples discussed in this book illustrate the diversity and multifaceted nature of these mechanisms, while seeking to outline possible avenues for establishing and operationalizing a responsive form of governance at a supra-local but subnational scale.

The scale debate is essentially dynamic, presuming continuous adjustment and territorial reconfiguration in response to new identities, social-political relationships and affiliations, leading to new 'modes and scale of governance' (Keating, 1997: 234) based on the 'relational dialectic among and between' (234) key social (and political) actors. It is these key actors, therefore, and their networks, that are seen as the pillars of the (variable) scalar arrangements for governance.

These debates raise important issues about relationships between economic change, the shape and impacts of institutions and the role of regions as expressions of territorial identities. The level of abstraction of debate, however, takes us far from the reality of change and conflict in particular cities and regions. There is also an underlying problem of the circularity of some of these arguments. We could summarize the debate thus: new economic scales need new scales of governance in order to negotiate new economic and territorial identities. It is not surprising therefore when Swyngeduw argues that actual mechanisms through which rescaling takes place remain vague. We do know, however, that complex processes of inter-action between economic, governance and cultural issues need to be disentangled to account for particular directions of change in specific regions.

The scale debate starts from understandings of profound economic changes. But questions of regional governance have also had substantial attention in political science. In the next section we examine various strands of this debate and focus in particular on the idea of a shift from govern-ment to governance, explanations of changing institutional patterns and the rise of network theories. The section concludes with an examination of how many of these debates combine into what is claimed to be the 'new regionalism'.

Multilevel and networked governance for regions

Debate in the early 1990s about a Europe of Regions and the growing complexity of institutional relationships behind European regional policy led to a search for new ways of understanding intergovernmental relation-ships. There seemed to be no clear hierarchy in these relationships. Nation states were not subservient to European institutions and subnational govern-ments sometimes had independent relationships with a European level. Additionally, in the 1980s, the growth of Europe-wide subnational govern-ment networks (such as the Eurocities network discussed in Chapter 3) made the institutional map of European government appear even more complex. In relation to European policy there was no fixed institutional hierarchy and new ways of explaining overlapping interests and responsibilities were needed. Several attempts were made to clarify the new relationships. Bennington and Harvey (1994) proposed a metaphor of 'spheres' to capture the sometimes overlapping interests of two or more layers of government and the role of new subnational networks. In the discussion of the conse-quences of further European integration arising from the Maastricht Treaty, Marks (1993) proposed an idea of 'multilevel governance' suggesting that relations between European, national, regional and local governments constitute a new form of politics. Whereas the idea of a Europe of Regions suggested a weakening of the national level, in this multilevel or multi-layered picture of European governance the nation state retains an influen-tial role. The extent to which the subnational level enjoys autonomy remains subject to debate (see John, 2000). Many commentators point to the con-tinuing role of the nation states in setting and controlling budgets.

The emphasis, however, is not on any one level but on relationships, often informal or beyond treaty commitments, between levels, and thus on flexibility (see also Aschauer, 2000). In some cases the national level may be the dominant player. For example, regional economic planning in Northern Ireland is driven by the desire of the British government to wean a dependent region off public expenditure (Gaffikin and Morrisey, 2001). In other cases, the multisector 'partnerships' set up to administer European regional funds can appear to dominate regional governance. The international level of European governance is not fixed but a terrain of conflict between the member states, the commission and other institutions such as the quasi-autonomous Committee of Localities and Regions (COR) established in the early 1990s.

Which level of government has the upper hand cannot be assumed and varies across European experience. However, the idea of multilevel governance presents challenges to understandings of political institutions and processes. A further challenge comes from the increasing involvement of business and voluntary associations in government. This has happened at a number of levels. Corporate representation and influence in Brussels has always accompanied the location there of the European bureaucracy and Parliament. The associated corporatist politics interested academics in the 1970s and 1980s. More recently, the involvement of business and voluntary associations in regional development partnerships demands new perspectives on European politics.

Regions and political identities

Keating (1997) uses the term regionalism to refer to the emergence of overtly political regional pressures across Europe. Regional political identities can be said to have been particularly strong at various periods in recent European history, and point to the fact that regions are not merely administrative or planning areas. A sense of belonging and thus of stakeholding in the region's fortune among policy makers represents an important factor in regionalization (Aschauer, 2000). As a political movement, regionalism has much variety and, frequently, deep roots in historical cultures, often predating the Europe of nation states which took its shape by the end of the nineteenth century. Regional claims have been marked by violence in Brittany in the 1970s, in the Basque country and in Corsica until today. In some cases strong cultural identities have been built into governance institutions. The new Spanish constitution of the 1970s allowed regional demands to be expressed in new institutional relationships within the Spanish state. The UK, more recently, has experimented with elected regional institutions in the historic nations of Scotland and Wales. The rise of the Northern League in Italy exemplifies many of the themes about the rise of regionalism and relations between economic and political pressures. The League has strong roots in the industrial districts of the north (Wild, 1997).

Wild argues that the League is 'the self representation of industrial districts in their search for regional power' (Wild, 1997: 99). The Northern League has self-consciously attempted to manipulate history and territorial imagery to create a new region distinct from the rest of the Italian State (see Agnew and Brusa, 1999). The region of Padania stretching from the Alps to Umbria and Piedmont to Trieste exists not as a territorial entity but a political rallying point. However, Giordano (2001) looks at the electoral record of the League and notes that the party seems unable to spread outside its original heartlands and into this greater Padania. Originally, the League based its claims on locality. Trying to invent a new Padania region however has been less successful. Another lesson from the case of the Northern League is that regionalism develops in relation to nation states. The most successful period for the League was during debates about a federal Italy, but failure at national level brought forward new regionalist strategies. Making regional institutions involves struggles between national and local levels. Indeed it is the history of nation state making in Europe which provides the very different series of contexts in which new regional claims emerge. And these claims have to be located in specific stories arising from the 'distinct spatial, political and historical "geometry" of local or regional territories' which provide the collective support for regional claims (Marinetto, 2001: 320). In the case of Scotland for example, it was the nationalist politics of the 1980s, interactions between MPs, trade unions and party activists, and the history of opposition to the Conservative national government which laid the conditions for Labour's devolution project in 1997.

It is not just regions challenging the nation state. Reacting to the over-optimistic claims for a Europe of Regions at the beginning of the 1990s, Le Galès suggested that European governance is marked by a range of increasingly prominent regions and cities (1998: 267). Urban politics has rivalled regional politics in many parts of Europe. Some extravagant claims are made for the new role of cities in European governance and some see a revival of European city networks on the scale of the Hanseatic League. Such claims are echoed in the US by Pierce (1993) 'Across America and across the globe, citistates are emerging as a critical focus of economic activity, of governance, of social organization for the 1990s and the century to come.'

In few, possibly unique cases, it is possible to see signs of emerging city states. For example, in the case of London, there are questions about the compatibility of world city and national interests (Taylor, 1997). The Mayor of London claims that the city is being weakened by contributing too much of its resources to the national exchequer. However, just as with claims about a Europe of Regions the city state idea is at least premature. We can certainly agree with Keating that European governance is developing asymmetrically (Keating, 1997).

In this context, an interesting question is how far individual cities and regions have developed the capacity to assert their identities and act in complex institutional settings. Thinking about how the capacity to act may be developed shifts the focus from formal institutions and on to the factors that drive regional aspirations and policy agendas. For the Italian Northern League, for example, there is a question of autonomy of economic policy making. Regionalism can also represent progressive agendas. Keil (1998), for example, emphasizes both struggles over quality of life on a regional scale and the potential for alternative formulations of environmental issues and local action (1998: 631). This form of regional politics involves action from church, labour and environmental groups. Progressive politics thus may include city-region confederations, community-based regionalism and cross class, multiracial coalitions. Such progressive regionalism contrasts with, for example, the Northern League and its desire to protect its economy against demands from the south. But this economic push behind new city and regional governments may reflect a wider reality in contemporary Europe.

Growth politics and regionalism

While progressive ambitions may motivate some regional leaders or city mayors, economic competitiveness remains the major driving force behind regionalism. Closer European integration, the creation of the single market and single currency have heightened competition between cities and regions (see Cheshire and Gordon, 1995; Newman and Thornley, 1996; Jensen-Butler *et al.*, 1997). At city level, the place competition engendered by economic integration has brought new understandings of the nature of city politics. The 'entrepreneurial city' was identified at the end of the 1980s as a response to a new competitive Europe. These cities both had strong mayors as figureheads but also new and close relationships between public and private sectors (Parkinson *et al.*, 1992).

Several bodies of theory sought to explain these developments and, in addition to explaining new forms of city and regional politics, new perspectives have been drawn on to develop models of institutional development which Europe's cities and regions should emulate. Research from the US seemed to offer plausible accounts of how successful economic governance is put together in the new competitive Europe. The pre-existing US model of the city as a growth machine (Molotch, 1976) had much to commend it to European researchers. The newly competitive, boosterist cities which appeared in the mid-1980s – Hamburg, Montpellier – encouraged comparison with assertive and independent US cities. The theories of growth coalitions emphasizing public–private cooperation in pursuit of growth seemed to work in the new spatial politics in Europe. In Molotch's model (Molotch, 1980; Logan and Molotch, 1987) those political and business groups with interest in the locality get together to promote growth and dominate the politics of cities. In European competition researchers could identify the political and other leaders behind city marketing and promotion. A similar conception of the business-centred regime (Stone *et al.*, 1991) focused attention on the crucial alliance of business leaders and political power in order to drive forward urban development. Keating proposed a similar idea of a 'development coalition' that could be applied at the regional scale. There are now almost twenty years of comparative studies built upon these basic concepts. The approaches seem to have stood up well to critics and give a fairly robust account of the politics of development (Lauria, 1997; Jonas and Wilson, 1999). Some (e.g. Painter, 1995) have tried to tie the local regime into the much broader regulationist account of economic change, others pointed out the important differences in translating essentially US concepts to European experience and, in many ways, the European city just did not fit (Harding, 1994). National politics, and political parties in particular, gave a different context to European city politics. In Berlin, for example, Strom (2001) finds all the signs of business centred growth but not the institutions predicted by US models. But recent assessments (Lauria, 1997; Jonas and Wilson, 1999) continue to recognize the importance of growth politics in shaping development but with varying suggestions for how such analyses might be improved or made more sensitive to local conditions. Another line of criticism of the growth coalition model is whether it works. Logan *et al.* (1997) is sceptical about any relationship between growth politics and economic performance. Pro-growth politics may squeeze out other redistributive or environmental agendas and international ambitions may not result in moves up the urban and regional hierarchy. Organizational, legal and technical difficulties can stand in the way (Beauregard and Pierre, 2000). The organizational resources to sustain pro-growth policies may not exist and political difficulties can result from the tension between international competitive goals and meeting local needs. Nonetheless, some commentators suggest that the necessary organizational capacities can be developed and adopted by cities with competitive ambitions (van den Berg *et al.*, 1997) and argue that cities wanting to pursue major projects require strategic networks, leadership, vision and strategy as well as needing to secure political and societal support. This new style of management seems to van den Berg as necessary 'to create and maintain organising capacity is really the key to dealing with the challenges that face cities today' (271).

Networking-defined regions

The idea that combinations of forces lie behind the decisions that govern the development of cities and regions reflects a much broader shift in academic discourse. The dominant model of social organization underlying much urban and regional analysis is that of networks. Keating (1997) emphasizes the importance of understanding relations between actors at new scales. The identification of intergovernmental and public–private networks has been an important part of both those studies we examined of the economic geography of regions, and work on changing relationships between political institutions. However, a weakness in the network approach is a tendency to identify and map network connections between governmental levels or between public and private sectors but without adequate attention to the particular ways in which different interests succeed in their objectives and manage to bring others in line with their objectives. Some networks and some network members are more powerful than others, and underlying power relationships need to be exposed in addition to just mapping interconnecting networks. The descriptive nature of work that applies network models is a serious weakness. Network theory needs a theory of power to explain which interests dominate and which outcomes appear, but such precision becomes increasingly difficult. Institutional hierarchies have been displaced by network forms and even in the liberal Europe of economic integration, corporate elites may not always have a dominant role in negotiating with local, regional, national and supra-national levels. Identifying the exercise of political power in cities and regions requires detailed studies and, perhaps in response, the case study form of analysis has proliferated (for example, Newman and Thornley, 1996; Healey *et al.*, 1997; Jouve and Lefèvre, 1999).

The case study perspective also responds to the arguments associated with 'institutional' theories in social science. This group of theoretical approaches argues that institutions make a difference in governance and can constrain public policy choices. Rather than looking at radical change these theorists focus on incremental developments which can be seen to be path dependent (see Pierson, 2000). Previous choices and institutional forms impact on the present and it is important to see how historical and cultural factors shape current decisions. Given this view we would expect different cities and regions to respond to competition or other economic and environmental pressures in differing ways. Critics of the institutionalist perspective, however, argue that there is still a difficulty in identifying the external or internal forces which bring about change and result in policy shifts (Gorges, 2001).

This debate about the relative importance of institutional and non-institutional factors is clearly important. The question of how non-institutional factors interact with formal political and administrative processes echoes many of the claims about a major shift from 'government' to 'governance' in political science (see John, 2001). If the concept of government fixes our gaze on governing hierarchies and formal decision making, governance points to horizontal networks of influence, intergovernmental cooperation, the blurring of public and private boundaries in decision making and the arrival of new private management techniques inside public bodies.

The concept of governance has also come to mean certain positive features of change, such as permeability between organizations and building the capacity to get things done. Governance can be seen, not as a means of control but as an 'attempt to manage and regulate difference' (Kearns and Paddison, 2000: 847). The concept is applied at both city and regional levels, and one of the challenges of analysis of contemporary governance has become the seeking out of shared interests in regional cooperation. From the bottom up new governance institutions solve collective action problems for groups of local governments. If hierarchical strategies no longer seem appropriate then effective governance is likely to be sought through network models. Research drawing on these foundations therefore examines the development of urban policy over time, the development of trust between members of networks and the important role of leadership in coordinating the new governance.

However, the claims of new governance and new networked forms of decision making may be over ambitious. While it is clear that the transformation of subnational European governments and a 'Europeanisation' of local and regional politics (John, 2000) is under way, it is perhaps too soon to make substantial claims about the outcomes of the new governance and its 'evolutionary advantage' (Jessop, 1998: 32). A second issue is that while cities and regions have clearly become the focus for new policies and initiatives, it is not always at these levels that solutions to problems are developed. Thus, saying that urban governance is multilevel may mean that non-urban levels of government (national, international) are involved. As we shall see in Chapter 3, the experience of European regional policy can be very much a top-down experience. A third limit to the claims of governance is that urban and regional levels are institutionally underdeveloped in many states and there may be no capacity to develop governance or new regionalism. In many cities traditions of clientelism stand in the way of new politics (see Seixas, 1999). Elsewhere, urban and regional levels run into conflict. Jouve and Lefèvre (1999) argue that the entrepreneurial mayors, who emerged in France in the 1980s, may not be as independent of other political institutions as they would have us believe. Existing institutions may resist new network or entrepreneurial forms. Thus urban and regional identities and cultures need to be brought into our understanding of how governance actually works. Institutional change at city or regional level often struggles to keep pace with economic drivers. In some ways this is not a new issue for city and regional government. The problem of 'boundedness' between economic and administrative territories is a long standing one in regional science. But solutions to institutional issues are being sought in new ways. The solution is now not necessarily hierarchical. There are possibilities of cooperative networks managing the new economic spaces of the European economy. That the development of the capacity to act happens unevenly should not come as a surprise. What this does is to force us to examine in detail how, in particular regional and urban areas, international, national, regional, and local forces interact. Research grounded in the new vocabulary of networks, institutions and governance is clearly moving in this direction.

These perspectives also have a normative dimension. Networking has impacts on longer term patterns of cooperation and in building institutional capacities. Some studies (van den Berg *et al.*, 1997) aimed to identify the specific institutional characteristics of success which could be transferred to less successful cities and regions in order to improve their competitive position. This 'organizing capacity' develops over time and emerges from core components such as leadership and strategic network formation as well as broader societal support. A similar position is taken by Healey *et al.* (1997) who see an outcome of the dialogue between multiple actors as forming over time 'institutional capital' which can be drawn on to form yet new alliances (1997: 23). But the search for the essential qualities of successful governance can prove elusive. In the US, Foster (2001), for example, tries to isolate 'regional capital' as a set of factors which support better economic performance through effective regional governance. Unfortunately there is no clear relationship between the stores of capital – historical, legal, structural, socioeconomic, developmental, civic, corporate and political – and regional economic outcomes.

For some commentators new, networked forms of governance open up new progressive possibilities for urban and regional governance. Amin and Graham (1997), for example, see progressive aspects of 'reflexive' networks, that is those that consciously adapt to external challenges. Rather than following hierarchically imposed rules and policies, new strategic directions for cities and regions could emerge from within. Such new political capacities may compensate for the widespread disaffection from traditional politics (see Clark and Hoffman-Martintot, 1999). Citizens may become more attached to governing regimes that can deliver, especially the 'non-material' public goods sought by the new middle class. New networked forms of urban and regional governance may be better at competing with other cities and regions and at delivering some types of local services. These sorts of normative claims have now much in common with the ideas of the 'new regionalism'.

The 'new regionalism'

The 'new regionalism', which emerged into academic and policy debate in the 1990s, picks up many of the key issues raised earlier in this chapter. Arguments about the new regionalism have a starting point in the assumption of much greater importance for regional economies. This view has several significant proponents on both sides of the Atlantic. Scott (1998), Cooke and Morgan (1998) and Barnes and Ledebur (1997) elaborate different versions of the case for regionalized economies, highlighting their specific qualities and identifying the policy benefits of managing on a regional scale. The region has emerged as a vital economic unit. For Barnes and Ledebur, 'Our argument is that the "national economy" is spatially differentiated and that local economic regions are the crucial units for focusing analysis and policy' (1997: 3).

They argue that the idea of the national economy is no longer relevant and that economic management should focus on interconnected regions in a global context. What also makes the regional scale important are institutional factors, for example the complex ties between central business districts, 'edge' cities, and suburbs and the networks of governance that manage them. The regional mix of institutions is argued to impact on economic performance. At base then the new regionalism asks us to shift our focus from national to regional economies and to the linkages between economic performance and institutional context. It is, however, difficult to generalize about the new regionalism beyond these basic points, as these views develop in different directions in the US and in Europe.

In the US, the new regionalism is not simply an argument of academics revising economic geography. The concept also refers to the numerous cities where active mayors have attempted to promote more city–suburban cooperation (see Orfield, 1997). These debates about city-regional cooperation revive long-standing concerns about the interdependence of cities and suburbs. Proponents of the new regionalism seek to prove the mutual economic benefits of cooperation on infrastructure investment, environmental planning and service management. Swanstrom (1996) puts the case for regionalism, emphasizing cooperation on economic but also equity and other grounds. Arguments for city–suburban cooperation stress the economic costs of separation between city and suburb and the mutual benefits of cooperation. Cooperation is essential for the containment of sprawl and promotion of 'smart growth' (Ross and Levine, 2000: 319). The economic argument which emphasizes effective relations between businesses and between business and communities at regional scale is thus joined to arguments about equity and environmental performance across regional economies.

The US debate draws attention to the few cases of region-wide government institutions and regional policy but also to the majority of city regions where such cooperation is the exception. Regional level institutions continue to be in the minority in the US. Only in exceptional cases such as Portland, Oregon, does regional cooperation produce apparently workable regional policy. The Portland case reveals a number of overlapping regional scales. City–suburban cooperation and the containment of sprawl are governed by an Urban Growth Boundary. Beyond the city, smaller towns cooperate on the planning of transportation links across state and international boundaries. These initiatives are exceptional and driven by federal programmes to foster metropolitan cooperation (for example, the Intermodal Surface Transportation and Efficiency Act, ISTEA, in the early 1990s and NAFTA (T21) transport subsidy). Regional cooperation in Cascadia (a notional region stretching from Eugene, Oregon, to Vancouver, British Columbia) originated in environmental lobbies in the 1980s but the idea was later incorporated into the economic marketing of the region (Harvey, 1998). So far the Portland and Cascadia cases are exceptional. The US new regionalism is more a set of ideas about economic and environmental cooperation than current reality. The few examples of formal governmental cooperation at regional scale and these exceptional cases are outweighed by examples of minimal cooperation and intra-regional conflict (Savitch and Vogel, 1996; Ross and Levine, 2000).

In Europe, the new regionalism suggests that there are lessons to be learned from successful regions. In many European states regional economic planning has a long history but it is competition and the search for competitive advantage at regional scale that lie behind the search for the institutional lessons that can be learned from successful regions. For Amin (1999) there are differences in the character and density of both politics and civil society at regional scale. Successful regions seem to have specific characters. Thus, 'Many of the prosperous regions of Europe are also regions of participatory politics, active citizenship, civic pride, and intense institutionalisation of collective interest – of society brought back into the art of governance' (1999: 373)

The 'prosperous regions' cited include the Italian industrial districts that we discussed earlier and Baden-Württemberg. Proponents of the new regionalism argue that having the right institutions creates economic advantage. Political forms, including elected regional government, seem to be important factors in making regions competitive. However, there are obvious difficulties in transferring lessons and a single model of change is unlikely to fit all cases. For example, Harding argues that elected government does not make a difference to regional performance (Harding *et al.*, 1996). National differences and the context provided by national, regional and urban policies will clearly impact on the emergence of new regional institutions. A broader critique of the new regionalism comes from Lovering (1999) and Tomaney and Ward (2000), who argue that there is little substance to the claims. They suggest that while it may suit some local government leaders or national governments or business leaders to argue for regional governance and a new regionalism, in fact there is no single, inevitable model of regional development either economically or institutionally. Neither does new regional policy necessarily benefit all localities. The pursuit of regional advantage may well force regional leaders to prioritize economic over social objectives and the needs of less advanced localities within regions.

This scepticism about the claims of the new regionalism should come as no surprise given the highly differentiated institutions of European states and the variation in economic circumstances that we go on to examine in Chapters 3 and 4. The claims of regionalists in the US and Europe (Pastor *et al.*, 2000; Amin, 1999) may well be overstated. However, these debates do point to important factors and processes in the development of regional economies and the institutions of regional governance. Perhaps the most important issue that becomes clear in the debates about new regionalism is understanding the ways in which existing institutional structures shape the emergence of new institutions at regional scale. Regions are shaped by both higher and lower levels and by the legacies of previous attempts to govern at this scale. Regional institutions include both formal tiers of government and informal coalitions of interests. Sweeping claims about a single model of regional governance run up against the variation of European experience. Against this background in the first part of Chapter 3 we examine the economic geography of contemporary Europe and what seem to be fundamental differences between and within Europe's regions.

Towards a new understanding of 'regions'?

In Europe, for the most part, the driving force behind the new regionalism has been economic competition. But there are other perspectives on the reasons for city-regional cooperation. Swanstrom (1996) and Pastor *et al.* (2000) in the US promote the equity environmental and economic objectives of 'growing together' through regional cooperation. The European regionalists and city regionalists (COR, Campaign for the English Regions) argue a democratic case for devolution and policy making at the regional level. Ache (2000) argues that local actors need to mobilize a range of interests around new visions of city-regional space, and Blotevogel (2000) emphasizes the changing nature of regions moving from fixed territory (spatial container) to regional groupings of specialized production clusters, which requires flexible policy responses. Strong arguments are thus mounted for regional governance, but evidence to support these normative claims is inconclusive.

A single vision of regional development and institution building won't work. But some of the theoretical developments we have reviewed in this chapter do offer potentially useful insights and suggest why we should expect difference rather than uniformity in city and regional institutions. Economic change impacts on city and regional institutions. The debate about rescaling, while offering a persuasive account of broad trends and pointing to a shift from centralized to polycentric decision making, was less clear about actual mechanisms and impacts. Economic and political contingencies make for a range of local responses to general processes. It is through case studies that we can begin to understand the interactions of these forces. We drew a similar conclusion from the discussion of multilevel and networked governance and the change from government to governance. National differences intervene, as do cultural factors and the histories of relationships between institutions.

Perhaps the most significant shift in urban and regional theory that we can detect in these debates is the fact that we no longer expect to find simple accounts of the relationships between city and regional development and policy and certainly no longer expect a 'one size fits all' model for city-regional governance. We do, however, expect the changing governance of cities and regions to be shaped by the interaction of economic and institutional factors mediated through political, cultural and other contextual factors. This perspective leads us towards an answer to the question that recurred in the early part of the chapter – how far are cities and regions in control of their fate in a competitive European and global economy? One impact of the work we have reviewed is that it has become difficult to claim simple relationships between economic imperatives and institutional responses. Policy choices will be forged out of economic circumstances and the articulation of complex configurations of institutional, political and cultural forces. No one scale of governance stands in a privileged position and we can only identify these relationships, not through theoretical principles, but through close examination of actual cases. In Chapters 3 and 4 we explore the changing economic context for European regional development and the evolving context of European policy and institutions. Chapter 4 shifts attention to differing national contexts and how fundamental constitutional differences impact on city-regional governance. These chapters then provide the background to the detailed studies of the different ways in which economic and institutional forces interact in the English and German cases.

3 European regions and regional policy

Introduction: regions, cities and policy efficacy

In Chapter 2, we encountered differing interpretations of changes in the relationships between the economy, territory and institutions of governance. Some of the claims of a 'new regionalism' are overstated. However, there is certainly widespread agreement about a growing diversity in the space economy and evidence of change in the management of cities and regions (Blotevogel, 2000). These changes are not all moving in the same direction and it is clear from the debates we examined that any account of regional and city-regional change needs to take in a broad range of economic, political and cultural factors that can interpret change at European, national, regional and local scales; and this is what is undertaken in the later chapters. This chapter looks at the European scale and examines two themes. The first is the background to evidence of spatial economic dynamism. The process highlighted here is growing regional economic heterogeneity, including between cities and regions, reflecting differentiation within existing formal regions. Such localized dynamism, of course, presents challenges to conventional, standardized regional policies which have little concern for flexibility and sensitivity to smaller scale and local opportunities. Heterogeneous economic regions contrast with those that seem to be structurally more homogeneous economic entities, such as rural areas which have been the traditional focus of regional interest (Stiens, 2000). The large, core regions have development potential extending well beyond individual local economies. Examples are the main capital cities, e.g. London, Paris, Berlin, or the main economic heartlands (Ruhr, Midlands, Randstad Holland, Baden-Württemberg), or the newly identified European Metropolitan Regions in Germany. At the same time, however, such strong centres may follow their own economic agendas on the basis of functional links and exchanges between them. Such functional networks would operate at greater speed and thus wider scale (Stiens, 2000) and thus be likely to bypass other, less well developed regions in between the contact points, even if they were closer geographically. A substantial regional challenge occurs in those regions showing considerable uneven development processes between their cities and the rest of the region. We argue that such localized development challenges require policies to become more spatially diverse, based on the needs and opportunities of new growth centres and other areas.

The second theme is the response of policy makers, at European scale, to divergence and diversity within and between regions. The second part of the chapter tracks the development of European policy and some fundamental shifts of direction as the EC begins to focus on urban areas in addition to its longer standing involvement at regional level, *inter alia* in the pursuit of greater policy efficacy to control costs in an expanding Union. The greater urban focus is also in recognition of the new role of the region as an 'important arena of collective decision-making processes' (Voelzkow, 2000: 513) and thus the importance of a distinct identity. Cities offer the common platform for such negotiations in 'their' regions. The degree of underlying social and economic heterogeneity will, in this context, be crucial for the possibility of identifying and developing such common purpose and identity in the interest of effective planning and policy making.

Economic heterogeneity and homogeneity of regions: localization of economic regions

This section outlines the distinct and changing inequalities in economic development throughout the EU, and the differing roles of cities and regions within that process. A contrast is drawn between homogeneous economic regions which should show few localized differences in development, indicating little dynamism within that region for further differentiation, and, at the other end of the scale, 'volatile regions', those that exist as regions merely on paper and depend on collaboration between dominant local urban centres. Comparing Europe's regions for different economic indicators provides an opportunity to look at the reality of the formal regionalism in different parts of the European Union.

There is a, 'recognition of the existence of significant disparities in the trajectories and development of different nations and of the mosaics of cities and regions that make them up' (Dunford, 1997: 90).

Disaggregation on a smaller urban-regional level provides a better basis to assess the geographically differing economic development of regions. The base of regional resources circumscribes its development possibilities. Resources include (following our discussion of 'soft' infrastructure in the previous chapter) social, cultural and institutional factors, inter-firm relationships and networking strategies, regional systems of innovation, mobilization of finance for economic development, relationship between government and industry, and the organization of powers between government tiers (see Dunford, 1997: 96). Policies and subsidies add to these resources (Dunford, 1997: 96) and an area's capacity to utilize these resources, including the formulation and implementation of policies, to achieve high productivity and growth is the foundation of developmental dynamism.

(a)

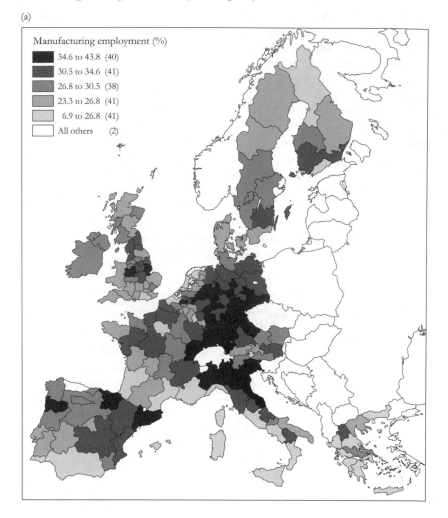

Map 3.1a Sectoral employment in the EU in 1996 – manufacturing.

Data source: Eurostats.

Overcoming these differences in spatial economic potential has been the main task of EU regional development funds that we examine in more detail in the second part of the chapter. Despite the ERDF's operation for some 30 years, differences in development opportunities and potentials remain, increased by the EU's expansion and thus growing diversity in national and regional economic structures. The planned eastward expansion into Central Europe is set to exacerbate the range and depth of inequality and thus the challenge to integration (we return to this issue later in the chapter).

(b)

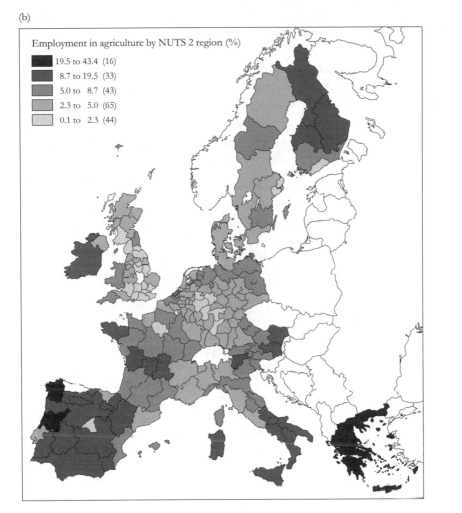

Employment in agriculture by NUTS 2 region (%)

19.5 to 43.4 (16)
8.7 to 19.5 (33)
5.0 to 8.7 (43)
2.3 to 5.0 (65)
0.1 to 2.3 (44)

Map 3.1b Sectoral employment in the EU in 1996 – agriculture.

Data source: Eurostats.

In addition, the use of established statistical regions as fixed, standard terri
torial units for dispensing structural policy (see Maps 3.1a and b) has
increasingly been challenged by a more diverse smaller-scale economic geography. Growing spatial differentiation in economic development capacity
has highlighted the role of urban areas as 'growth centres' or 'innovation
centres', for example. Thus, there is evidence of a new, more diverse and
volatile economic geography of the EU. These differences may question the
appropriateness and effectiveness of conventional regional policies with their

fixed territorial structures and inability to respond flexibly and sufficiently quickly to changing economic geographies.

Economic differentiation and inequality in the European Union: evidence of a new regionalization across administrative boundaries

There are strong indications that differences in social and economic development across the EU continue to be considerable and are likely to increase with the accession of former socialist countries. In particular, the pattern of these variations will become smaller scale and will further highlight the renewed importance of the 'region' particulary in eastern Europe (see also Bachtler and Downes, 1999, 2000). The European Commission's own assessment of the development of inequality therefore emphasizes the importance of scale when evaluating inequalities. Depending on whether one chooses national or subregional territories, the Commission says, 'over the last 10 years, differences can be said to have either decreased, remained unchanged or increased, depending on the scale used for the assessment' (EC, 1999b: The European Union: Cohesion and Disparities, www. inforegio.cec.ei.int/wbover/overcon.oco2a_en.htm accessed 4 Nov. 1999, p. 1). Thus, at the most aggregate national level (NUTS 0), inequalities in economic performance (GDP per capita) appear to have been reduced between the relatively poorest (Mediterranean) countries and the EU average in the 1990s. In Ireland, another peripheral state, for instance, economic performance (GDP) improved from 64 per cent of the EU average in 1983 to 80 per cent in 1993 and 90 per cent in 1995 (EC, 1999a: 2).

Using comparative differences in GDP and a ranking of regions (NUTS 1), it appears that little has changed since the mid-1980s, with eight of the ten richest regions included in the top ten in both 1986 and 1996 (EC, 1999a). A more detailed picture of changes and emerging patterns of inequality can be obtained from measuring the overall differences between the EU average and relative performance of all regions in relation to that. The standard deviation provides such information in a more comprehensive and accurate way than focusing on the top best and bottom worst only. A decrease would indicate relative convergence among all regions, although corners of poverty may continue to exist, somewhat obscured by greater improvements elsewhere (Table 3.1). There are some indications of a slight narrowing in discrepancies by three percentage points at the bottom of the table in favour of the least well performing regions, while at the top, differences have increased (EC, 1999a). This reflects the success of some growth regions when it comes to attracting new investment and participating in the top end of the locational business market.

The situation looks less convincing when taking the next lower spatial level, NUTS 2, of the official territorial hierarchy used as the basis of structural regional policies. Gaps in economic performance (GDP per capita)

Table 3.1 Disparities in economic output (GDP) between and within EU regions

State	Index of GDP per capita (EU15 = 100)		Regional disparity of GDP per capita (standard deviation)	
	1986	1996	1986	1996
Austria	103.2	112.3	24.7	28.6
Belgium	102.8	112.1	25.0	26.0
Denmark	112.1	119.3	–	–
France	109.8	103.2	27.8	29.0
Germany (unified)	–	108.3	–	30.2
(West only)	116.1	118.5	22.0	23.7
Greece	59.2	67.5	6.0	8.6
Ireland	60.8	96.6	–	–
Italy	100.4	102.7	25.2	27.2
Luxembourg	137.3	168.5	–	–
Netherlands	101.8	106.8	12.2	12.3
Portugal	55.1	70.5	16.2	13.1
Spain	69.8	78.7	13.7	16.8
Sweden	111.5	101.2	10.7	11.1
United Kingdom	98.6	99.8	19.6	18.5

Data source: EC (1999a).

between the 25 worst and best performing regions has, if anything, increased slightly from 50 to 52 per cent and 140 to 142 per cent respectively of the average EU GDP per capita between 1986 and 1996.

The considerable differences in economic capacity between the EU countries (see also Dunford, 1994) are highlighted further by the composition of sectoral employment (Map 3.1). Relatively high agricultural employment can be found in the outer, peripheral ring of the EU's regions, such as the Mediterranean countries. It is interesting that France can be included in this set of countries, despite its status as one of the leading western European economies. This suggests considerable internal variations between 'advanced' and 'less advanced' economic regions within the country. Such internal variations also emerge for Spain and Portugal, but in those cases the 'less advanced' regional economies (with primary sector employment including more than 6.4 per cent) appear to dominate. Only the two capital city regions, the restructuring old industrial region around Bilbao, and the Costa Brava with its intense tourism, show a lesser importance of agricultural employment. In the latter case, this corresponds, not very surprisingly, to a strong showing of service industries in these areas, mainly tourism and administration. It seems that in some regions, such as the coastal strip of Costa Brava, an essentially 'pre-industrial' structure has been followed immediately by a 'post-industrial', service-based pattern. Others, like Bilbao or Barcelona, appear to restructure successfully from an old industrial to a cultural-tourism city. The capital city regions always had a high proportion of administrative employment, increasingly dominating the respective economies. Figures also show the importance of an urban–rural contrast in economic development outside the tourism-focused coastal resort areas, pointing to the challenges of twin track regional development between urban core and rural peripheral regions.

Sectoral employment also reveals the interesting position of Germany as the still most industrial economy within the EU, with most of the country, including some pockets in the old industrial areas in southern eastern Germany (e.g. Saxony), showing consistently high values of more than 35 per cent secondary sector employment. This is a considerably higher share than in de-industrialized Britain, for instance. However, the difference between 'old industrial' and 'new industrial' regions, such as the Ruhr and Stuttgart (Baden-Württemberg) agglomerations respectively, needs to be taken into account as official figures merely refer to 'manufacturing' and do not allow us to distinguish between 'smoke stack' and other types of manufacturing. Similarly high manufacturing employment can be traced in a few regions in the industrial core areas of northern Italy (e.g. Veneto) and eastern Spain (Andalusia), and, if rather more localized, in France and Britain. They include the Nord-Pas de Calais and, despite their heavy de-industrialization during the 1980s, the Midlands and the North of England respectively. But there are considerable variations within these regions. One example is Yorkshire and the Humber, one of the 12 English regions, where the shift from an industrial to a post-industrial urban region has varied greatly in its force. This is illustrated by the two neighbouring cities of Leeds and Sheffield, both situated within the same formal region (see Chapter 6). While the former has recently acquired a vibrant image and has been successful as a new regional service centre, the latter is still stuck with its steel town image. Local characteristics and particularities have largely accounted for these differences and emphasized the importance of city regions vis-à-vis the much larger and more heterogeneous, conventional formal region.

Overall, indicating the continued economic importance of productive industries, though to a lesser extent in the form of smoke stack manufacturing, the 25 regions with the highest manufacturing share in employment achieved wealth creation 8 per cent above the EU average (EC, 1999a). Compared with the performance of the main centres of service activity, however, this is much less impressive. Regions dominated by tertiary industries not only show fewer variations in economic output, but also a generally much greater importance in the labour market with at least 59 per cent of employment provided by this sector throughout the EU regions. Not surprisingly, therefore, the 25 regions with the highest employment shares in services are also among the economically most successful with a GDP level of more than a quarter (27 per cent) above the EU average (EC, 1999a). While this points clearly to a 'post-industrial' economic stage, there are distinct variations within the official formal regions, which become apparent when comparing sectoral employment shares at national and regional levels.

In Britain, the higher than average (65 per cent of the EU average) tertiary employment recorded for some of England's northern regions is fundamentally caused by local success stories of economic restructuring, such as Leeds and Manchester respectively, rather than a general economic boom throughout the regions. Also, not all service sector jobs are economically high yielding. Many are low skill jobs in call centres which can be found typically in economically more depressed, low wage areas.

The variation underlying the structural economic composition of employment reflects the varying relative strengths and weaknesses of regions and thus their appeal to business investment. Interregional competition for such investment is the inevitable result. Increasingly, variations within regions also gain in importance. With growing sophistication and detail of strategic locational decision making shifting to a more local level, even small variations within regions matter and affect the overall economic performance of an area now and in the future. Variations include those between urban and rural parts of a defined, formal region. The degree of urbanization appears to be directly related to economic performance, measured by GDP per capita, emphasizing the importance of city regions as economic core areas. Thus, in the EU, regions (NUTS 2) characterized as 'urban' have a GDP of almost a quarter (22 per cent) above the EU average, jointly generating some 60 per cent of the EU's total GDP (EC, 1999a). Put differently, each of the ten regions with the highest level of wealth creation in the EU includes at least one major conurbation (EC, 1999a). From a point of view of economic development opportunity, therefore, the formal EU regions may portray too grainy a picture of the underlying economic geography and the differential roles of localities within them. There may also be continuous changes in the relative appeal of areas, to which the fixed formal regions cannot respond sufficiently quickly.

A more accurate picture of the differential economic performances is revealed up at the more detailed and locally sensitive subregional (NUTS 3) level. Figures here point to a growing discrepancy between better and worse per forming localities within the formal regions. Such differences suggest variations in policy requirements of greater spatial targeting than conventionally afforded at the more aggregate scale of the region, so that indigenous development potential can be addressed more effectively. At the same time, however, such smaller-scale targeting may further enhance underlying inequalities in development. For instance, capital cities emerge consistently as the main foci of new investment and innovation and thus prosperity (EC, 1999a), potentially increasing existing discrepancies into their more lagging, wider hinterlands. The example of Berlin highlights these difficulties while questioning established formal regions and their seemingly artificial boundaries (see Chapter 6). Levelling out such strong unevenness in development potential and pathways across, in this case monocentric, regions will be very difficult to achieve. General standardized measures at a larger spatial scale without adequate attention to sufficiently spatially detailed, locally varying development capacities offer little. Instead, the main agglomerations may well steam ahead, helping to improve the *aggregate* figures for their respective regions, but *without* genuinely benefiting the region as a whole. Also, these centres may establish their own economic, functional and institutional-political alliances, transcending the boundaries of formal regions, and, in doing so, tending to ignore their wider regions. Stiens (2000) describes this process as a widening of a grid made up of linear linkages between major urban centres, leaving ever larger areas to fall through the raster and remain confined to economic (and political) peripherality.

Evidence of regional disparities

A useful indication of a region's equipment with urban development potential is the degree of urbanization, which itself is a reflection of population density (see also BFLR, 1995). The European Commission uses three main categories of 'density' (EC, 1999a): 'dense' is defined as at least 500 people per sq. km, 'intermediate' as a density between 100 and 500 people per sq. km, and 'thinly populated' as under 100 people per sq. km. On this basis, nearly half (49 per cent) of the population lives in densely populated or 'urban' areas (see Map 3.2), and about a quarter each in 'intermediate' and 'sparsely populated' (rural) areas. Spatially, the urban areas are narrowly concentrated on a mere 3.5 per cent of the EU's territory (EC, 1999a). This suggests a correspondingly narrow territorial selectivity of innovation and investment and thus wealth creation. These localized centres of high development potential are concentrated on a narrow band through north-west Europe, linking the main urban centres stretching from Benelux via western Germany to northern Italy. This ribbon has also been referred to as the 'blue banana' (CEC, 1990). The other main agglomerations, the English Midlands and Paris, are relatively close to this band.

The nature of urban regions will affect the likely dispersion of growth and economic prosperity. Monocentric urban regions, dominated by one large conurbation, e.g. London, Paris and Berlin, are more likely to retain growth than polycentric regions based around networks of smaller cities, often associated with lower centrality (Stiens, 2000). Polycentric regions, such as Yorkshire and the Humber region in England, Saxony-Anhalt in eastern Germany (see Chapter 6) or Emilia-Romagna in northern Italy, suggest a greater likelihood of including the non-urban parts of the regions covered by the network, because of the lesser degree of domination by the cities. The rural–urban mix within regions thus seems important for the generation and distribution of economic development. Policy requirements in the two types of city regions may thus be different in their need to address specific local as well as wider regional interests and structural characteristics, not only to promote economic development, but also to facilitate greater harmonization of development prospects and people's economic welfare across and within the regions, if that continues to be a key policy aim. Policies would, for

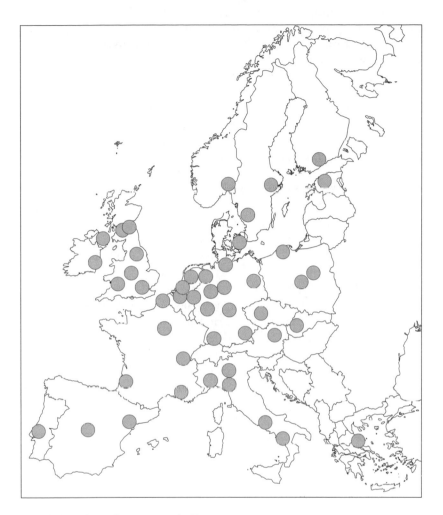

Map 3.2 Major urban centres in Europe.

instance, need to address the specific economic mix found not only in regions generally, but also (urban) localities in particular. Urbanization seems a necessary but, on its own, not sufficient condition for achieving good economic prospects. Rather, existing economic structures play a crucial role. On that basis, the rosiest prospects for regional economic development would be encountered by an urban region with a high content of new industries. Urban areas with more prominent old industrial structures would, by comparison, be in a less advantageous situation, possibly competing with less densely populated, and thus potentially less favourable semi-rural regions, but which possess the advantage of a new technology-based post-industrial economic structure. The least prospects would exist in those regions with few, if any, major urban centres, being dominated by a rural economy, or by a heavily declining manufacturing industry with few signs of new investment.

The main aim of EU regional policy is the achievement of cohesion and more egalitarian quality of life throughout the EU. One important indication of policy success would be a narrowing of the gap between the less well performing regions and the EU generally. In effect it has improved only little. This is despite the quite remarkable improvements in the four 'cohesion' countries of 1986, Spain, Portugal, Greece and, especially, Ireland (see below). The latter's economic output (GDP per capita) in relation to the EU average increased from 61 per cent in 1986 to 97 per cent in 1996 (EC, 1999a, Table 3). Similar, if not quite as strong, improvements were achieved by the other three countries. Some overall reduction in inequalities occurred as a result: the 25 regions at the bottom of the league increased their output by about 5 percentage points, although they still remain at below 75 per cent of the EU average. Much of this growth has been achieved through inward investment (EC, 1999a), encouraged by loans and grants and considerable EU supported pre-investment in infrastructure. The strong reliance on inward investment, naturally, will increase dependency on corporate decisions made elsewhere and give cities the key to regional (and national) development. This could lead to disinvestment elsewhere in more peripheral areas as a worst case scenario. There are, however, some unfavourable signs of a slowing pace in closing the GDP gap to the EU average, such as in Portugal (EC, 1999a). Market saturation, competition from low-cost eastern European countries and a weakening EU economy during the second half of the 1990s have been contributory factors in this process. Such slow-down may be expected for Ireland, which has generally achieved convergence with the EU average in economic output, following the phasing out of generous support under the EU regional policy.

Towards new regions?

Regional structures and economic development processes cross national borders. Clusters of regions with greater similarities with neighbours in other countries may be identified which deviate from nation-based territorial shapes. Using structural, economic and political common-alities, including geo-political, structural economic and urban characteristics as underlying criteria, the EC (EC, 1994) suggests a range of large-scale European regions. They are of a spatially generalized and informal nature and are thus less clear in their actual relevance to existing institutional and policy-making processes (Map 3.3).

Map 3.3 shows the various geographic locations of these 'super-regions'. Their common characteristic is a focus on the main urban areas or urban centres as foci of economic activity and sources of wider regional develop-

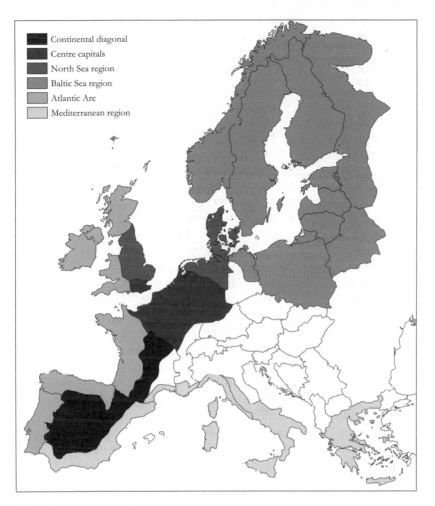

Map 3.3 Transnational concept regions.

ment, and thus reflect the changing focus of EU regional development policy discussed below. These urban centres are interconnected by functional 'corridors' which follow linkages between these centres. There is a long list of further European regions, eleven in total, with varying obvious strength and common ground, but with considerable political effort by individual cities and regions to be included in one of these large regions. Some of them, like the Atlantic Arc, appear to have little more in common than their relative geographic proximity to the Atlantic, including such different localities and regions as Lisbon, La Rochelle, Dublin, Belfast and Inverness. The economic context of these cities is fundamentally different, as are their culture, social-political and physical-geographic circumstances, making their proposed commonality somewhat difficult to see.

Using differences in GDP per capita growth rates for NUTS 2 regions during the 1980s, Rodriguez-Pose (1998) suggests a similarly scaled, if more realistic, regionalization across the EU, based on differences, and similarities, in underlying economic structures and development capacity, rather than political interests. On the basis of relative differences in developmental progress and dynamism, he distinguishes four main types of dynamism in regions (see also Dunford, 1994), reaching from the most dynamic 'capital and urban regions', via 'intermediate less dynamic regions' and 'peripheral less dynamic regions' to the least changing 'peripheral dynamic regions'. In addition, the capital city and urban financial centres may be distinguished, as they may hold ambitions to become 'global cities'. The rather coarse pattern offered by NUTS 2 regions becomes evident by the fact that the Frankfurt region, for instance, does not feature as a major financial centre, because its economy is largely 'hidden' in the quite large *Land* of Hesse. The main difference between the regional types is the degree of peripherality in relation to the main urban centres. On that basis, alliances between the strong and the weak may develop, if there is a mutual interest, but there may also be links between the strong cities, leaving the peripheries to seek similar allies to jointly overcome their difficult situation. In so doing, the emerging links may transcend rigid formal regional territories and, especially, national boundaries.

It becomes obvious that regional development processes have begun to develop their own dynamisms, based on interregional competition and difference rather than similarity. Such dynamism may include the emergence of new economic territories (regions) unaffected by existing administrative boundaries drawn on the basis of earlier structural economic or historic political circumstances. Such dynamism appears to be particularly vigorous in the main city regions with their appeal to investment and thus development input. 'The places that win are in many cases dynamic systems of cities and extended metropolitan city regions' (Rodriguez-Pose, 1998: 110).

This is mainly due to the concentration of highly skilled people in the main metropolis offering considerable locational advantage over other areas. Furthermore, metropolitan regions act as main switchboards in networks both electronically and in terms of interpersonal communication/collaboration. Not surprisingly, therefore, 'almost all of the ten leading regional economies in Europe are centred on metropolitan cities' (Rodriguez-Pose, 1998: 111), followed by the main industrial regions. With growing international competition, the skill factor will be of fundamental importance for high-cost countries such as those in the EU. It will be infinitely more difficult to compete purely on costs, such as in low-skill manufacturing activity.

An important question in relation to spatial inequalities within regions is the scope for achieving greater economic harmonization as the EU's main policy-making aim. Can there be a 'trickle down' from the main cities? The European Commission expects a two-stage development with (1) an initially widening gap in economically less developed regions between the cities as bridgeheads of new investment and development and the surrounding areas, and (2) the wider distribution of growth to the surrounding more rural hinterlands of the regions. This, however, requires continuous 'surplus' growth that cannot be accommodated within the cities' territories. Only then will such growth be exported to the surrounding hinterlands, establishing new economic interdependencies and a new 'second tier' subdivision of growth areas within the regions: the urban core, the dependent hinterland and the structurally more independent, if less affluent, rural periphery.

Further differentiation has become clear in recent years within city regions. And such disparities present a new set of problems for European as well as national and local policy makers. Kunzmann (1996) sketches the typical European city region which is constructed from a set of zones. The successful city region has a high level service centre, a zone attracting global tourists, new technology zones and high value residential enclaves. But it also has worn-out industrial areas and houses marginal populations. The interdependence of these areas with the city region – the airport zone and financial district, for example – presents substantial challenges for regional planning and horizontal and vertical cooperation between governments. As shown in the next section, differentiation within city regions, mounting problems of unemployment and social exclusion also draw the European Commission into a finer grain of 'regional' policy making.

European regional and urban policy

Successive European treaties have increased the penetration of European programmes into Europe's cities and regions. Those programmes impacting directly on cities and regions include those concerned with transport policy (for example, completing the Trans European Networks), environment policy (for example, the standardization of approaches to assessing environmental impacts since the Environmental Assessment Directive of 1985) and regional policy supported by a range of funds to support the weaker regions. The historical development of these European policy areas can be followed in several sources (e.g. Williams, 1996). This section traces some of the key themes of this development and concentrates on more recent shifts in focus. An important part of this development is the way in which regions and cities have been considered by the EU as objects of policy. Over the longer term we can discern a shift from a European policy focus on industrial sectors – the original post-war Coal and Steel Community as a basis of international cooperation – to a territorial and regional focus and, more recently, to a concern with the economic performance of Europe's city regions and the damaging effects of specifically urban problems on economic performance.

Despite increasingly systematic policy interventions since the 1970s, large disparities continue to exist both between and within Europe's regions. Convergence and cohesion in Europe have been long-standing policy themes. However, two issues in particular have come to dominate recent debates and developments in European policy. Expansion of the European Union has had profound influence on the scope of regional policy support since the close of the last century. The impact of the shift in focus arising from expansion into central Europe will be examined here. The second important issue is the increasing attention to policies that specifically focus on urban, as opposed to wider regional, issues. In the early years of European cooperation, policy makers faced major problems in old industrialized and coal producing regions across Germany, Belgium and northern France. There has, however, been a shift of focus from such perceptions of European regional problems to the problems of unemployment, congestion, and crime concentrated in some of Europe's urban areas. There is a third important issue that will be examined towards the end of this section. This is the shift in the institutional framework that is accompanying the changing European level policy, which applies both to the European scale where the roles of the Commission and other institutions change over time, and in the different ways in which European institutions interact with national and subnational governments. For Europe's cities and regions, there are therefore important issues of both policy and policy change and institutional adaptation to European policy. This dual focus of change echoes the discussion in Chapter 2 of rescaling and institutional innovation.

Growing 'Europeanization' of regional problems and regional policy

As was made clear in the earlier sections of this chapter the difficulties of finding a sound comparative base for regional statistics exposes the varied definition of regions in European states. However, while regional convergence has been at the heart of policy development, it would be unrealistic to see European policies towards regions merely as responses to statistical difference and relative need. The development of European institutions has from the start been a political process and the development of regional policy no less implicated in the machinations of political relationships between member states. The statutory commitment to pursue an 'ever closer union' includes greater harmonization of living standards and opportunities, such as found in most (traditional) national spatial planning policies. This common objective encouraged a continuous 'Europeanization' of initially largely nationally defined, if EU supported, regional policies, leading to greater EU control and definition of policy objectives and eligible projects.

At the ministerial discussion over the accession of the UK to the European Community in 1972 and with growing evidence of extensive decline in old industries, systematic funding for support for industrial areas in difficulties was agreed. Financial support for the UK's regions provided a means of directing subsidy to the UK to balance the substantial funds other states were to receive from subsidies to agriculture. This subsidy – the European Regional Development Fund – was based on *national* quotas with *nationally* defined regions as the beneficiaries. Systematic regional intervention thus started essentially as a national government exercise, with little regard to a state's federal or unitary organization. With subsequent accession of new member states, further adaptation of regional funding occurred. The accession of Greece in 1981 highlighted the extent of subsidy to better-off northern European states. Redistribution of the ERDF to include Greece reduced the proportion going to existing members but an overall increase to the budget at that time meant that no member state actually lost out and thus avoided political controversy and more fundamental debate about the scope of the EU. The potential political scale of EU membership costs, or financial benefits, became obvious in Margaret Thatcher's claim of an annual rebate on the UK's membership payments in the late 1980s to make up for perceived disadvantages in receiving regional aid.

Increasingly, during the 1980s, what was essentially a set of national funding arrangements, became gradually Europeanized. A significant turning point in this direction was the introduction of a series of new funding programmes in 1985. These programmes put forward common criteria against which individual projects were assessed, giving a specifically European dimension to the nature and distribution of funds. With the introduction of prescribed planning documents to support bids for funding in the European Community, regional policy slowly acquired more systematic forms as membership of the Community grew.

The next evolutionary step in regional policy came with the accession of Portugal and Spain in 1986. The inclusion of poorer countries challenged the existing distribution of regional funds. At this time member states were agreeing substantial changes to European treaties to create a single European market and the new treaty included articles on economic and social 'cohesion'. At the heart of this concept was the acknowledgement of disparities between rich and poor regions and a shared objective that these divisions should not widen as a result of closer economic integration. Regional policy was an important instrument in pursuing these aims. The scale of regional funding was increased, rising to about a quarter of the EU budget by 1993. An important feature of the Maastricht Treaty negotiations in the early 1990s was the proposal for a new Cohesion Fund to target support at the relatively poor peripheral states, Greece, Ireland, Portugal and Spain. During this period a new set of influences on European regions began to emerge. Environmental politics began to play a significant part in European policy at the end of the 1980s, especially following the election of thirty members of Green parties to the European Parliament. Accession of Sweden, Austria and Finland, countries operating high environmental standards, ensured a continued influence of environmental issues. In addition to this new environmentalism, the economic project of the single market was accompanied by European transport planning and the idea of TENS linking together states, and strategic transport investment. For cities and regions, location on new strategic road and rail links could have as profound an economic impact as the flow of regional development funds, because they could mean new, maintained or lost locational advantages. Inevitably such advantages accrue initially to those regions at the centre of transport links. Thus, while in the early 1990s a TGV network extending from Madrid to Berlin could be imagined, the Paris–London–Brussels link was beginning to operate. Nevertheless, transport investment has played a significant role in the development of the periphery. The EU invested in a new underground system for Athens, stations for Lisbon's Expo in 1998, in motorways across Greece and, looking further ahead, links to the east, to Moscow and Kiev by road and rail. The next main challenge to the EU's ways of making spatial policy is the envisaged eastward expansion to include the neighbouring Central European states. Specific policies, such as INTERREG and the establishment of Euroregions, have already been put in place (the role of INTERREG in cross-border planning is discussed in Chapter 4). The main discussions centre on the financial affordability of such expansion under the existing regime of regional policy, and thus the form and nature of required changes.

Against the background of greater Europeanization of structural policies, a growing and even more direct impact on regional policy developed through a series of initiatives concerned with spatial planning as a more direct form of centrally directed policy making.

European spatial planning: towards informal cooperation

At the beginning of the 1990s a series of reports from the European Commission began to address urban and regional issues in a comprehensive fashion. Europe 2000 (CEC, 1991) and Europe 2000+ (EC, 1994) identified core trends and repeated familiar overall themes of reducing disparities, while facilitating economic and social cohesion with the objective of creating a better balance between regions. In the latter document, the map of Europe was separated into several large 'regions' which, it was claimed, shared common problems (but see the discussion in Chapter 4) and which faced common challenges. The document outlined 'trend scenarios' for the super regions as a precursor of some potentially more developed regional policy statements. Some of these regions, such as the Baltic region, crossed the EU borders and prefigured the issues to be addressed in debates about accession of central European states. The Commission intended that Europe 2000+ should be followed up by a series of transnational studies, both to provide a better understanding of problems and to develop cross-national ideas about future developments. The Atlantic Arc area had its origins in

Brittany in the Atlantic Arc Commission which provided a legitimate bottom-up super-regional initiative as a model for Europe 2000. The French national planning agency, DATAR, actively promoted an Atlantic Arc as a counterweight to the identified core of the European economy in Italy/Germany/Benelux and southeast England (the so-called 'blue banana'). The Euroregions, defined in Europe 2000+, modified the original map of the Atlantic Arc but maintained the linkage from the south of Portugal to the islands off the north of Scotland. The area had support from the early round of INTERREG, a funding programme of the European Commission to support cross-border cooperation, but an important emerging theme was the influence of national government in what had been originally conceived as a devolved distribution of funds (Wise, 2000). In INTERREG II the Arc joined funding programmes aimed to support the new European Spatial Development Perspective. The, at first, EU-shaped regional initiative was thus now defined in boundaries set by the nation states, including more Spanish regions, and stretching further into France. The development of policy and programme funding in this area illustrates well the tension between regional aspiration and national power in European policy. The assertion of national power slowed the process of bringing forward local projects, and Wise (2000) contrasts the period of national controls with the successive period of vibrant cooperation between regional authorities, professional groups and civil society as expression of a shift from top-down, often technocratic, government, to more democratically controlled governance. In addition to the national–regional conflicts in conjunction with the development of the super-regional scale of governance, the history of cooperation at this scale also exposes the strong national differences in the approach to devolution and in the ability of national and regional interests to succeed in the competition for European funding. Where strong regional agencies already exist, in the case of the Atlantic Arc, for example, in Scotland, then there may be less interest in the super-regional routes to funding. Equally, strong national involvement in EU processes may weaken interest in this super-regional route, because it may mean less power. In short, the experience of the super regions exposes the asymmetrical development of governance institutions between local and national scales.

Member states had little interest in pursuing super-regional studies and plans. Equally they had only limited interest in the series of informal discussions between ministers on spatial planning which began in 1989. By 1994 the meetings had the status of informal councils of ministers and the Committee on Spatial Development produced a first draft European Spatial Development Perspective (ESDP). The origins of this initiative lie in concerns on the part of the German *Länder* that the Europe 2000 initiative would lead to the Commission producing a top-down European spatial plan, thus undermining their policy-making autonomy (Faludi, 1997).

The definitive version of the ESDP, adopted by the planning ministers in 1999, identifies the goals of cohesion and sustainable and balanced development. However, it remains an informal, consensus document with limited means of moving towards these objectives (Faludi, 2000). At the heart of the ESDP is a tension between the desire for balanced development but within a framework of globalization induced competition. The ESDP argues that stronger integration of European regions into the global economy benefits the competitiveness of the European economy and thus works towards the traditional goal of improving qualities of life. At the same time, the ESDP acknowledges interregional differences in competitiveness and thus opportunities, when it identifies leading regions as 'global economy integration zones'. 'The creation and enlargement of several dynamic global economy integration zones provides an important instrument for accelerating economic growth and job creation in the EU, particularly in regions currently regarded as structurally weak' (CSD, 1999: 20).

In earlier drafts and in the Europe 2000 studies, competition had also been associated with the negative impacts of congestion on environmental quality in the regions of the core of the European economy. The ESDP, however, stresses the economic benefits of competition. Jensen and Richardson (1999) identify two dissenting voices. First, states on the southern periphery which may be prepared to go along with the idea of catching up core regions as long as the flow of funds for infrastructure projects continues. However, as shown later, the recent changes to the Structural Funds and the likely shifts which would follow the accession of central European states, might well undermine enthusiasm for the idea of urban and regional competition. Problems can also be foreseen on the northern periphery. One argument there is that while the ESDP favours support for globally competitive cities and regions, most of this part of Europe is not urbanized. The ESDP makes mention of urban–rural relationships but rural regions do not figure as 'global economy integration zones'. A further political issue lies in the varying responsiveness of differently developed local democratic systems to the imposition of a large-scale regional model of the future of European space.

The ESDP marks a significant development in European urban and regional policy, but the document's propositions are weak and resources that could support European spatial planning are limited. The ESDP has few ways of tying its model of competitive regions to national and regional plans. Some states (Netherlands, Denmark, Germany) have adapted national planning documents, albeit with varying detail. Ireland published a new National Plan 2000–2006, that follows ESDP principles in promoting balanced development between the dynamic Dublin city region and the rest of the country. Others are writing the ESDP into regional plans and both regional and national plans will be used to support claims for European funding. The only direct line of budgetary support for the ideas of the ESDP exists through the INTERREG programme of funding for cross-border planning and development. For the period from 2000 to 2006 this programme offers 4,875 million ecu to selected projects. The INTERREG IIIC supports cooperation and exchanges of experience and best practice in the priority Objective 1 and 2 areas defined by the ERDF. The potential areas of cooperation are broadly defined, but the administrative arrangements are complex. The programme defines large European regions (the UK being within the North West Europe Area that includes Luxembourg, Belgium, Ireland and parts of France, Germany and the Netherlands), but also allows for cooperation beyond these regional borders.

From peripheral regions to city regions: new urban focus and the URBAN programme

In the studies for the Europe 2000 and Europe 2000+ projects, urban Europe came to the fore (see CEC, 1992; EC, 1994). These studies began to build an argument that Europe's cities, rather than its industrialized and lagging regions, had greater relevance to the process of economic and monetary integration. In addition, the review of Structural Funds post-Maastricht increased the Commission's capacity to launch new initiatives. Most of these now had an urban focus. The lessons from successful cities (CEC, 1992) suggested that local institutional capacity was a significant factor in promoting endogenous growth, shifting and extending the focus from conventional physical infrastructure projects to systems of governance. An urban policy focus thus began to make sense in terms of reshaping the European economy and moving towards the goal of internal social and economic convergence and integration. This was made explicit in the launch of the URBAN programme in 1994 with 85 individual programmes focused on infrastructure and measures to tackle high unemployment.

Assessing how effective this targeted urban policy could be brings us back to some of the basic concerns underlying regional policy. Northern cities had more experience in dealing with Brussels, of gathering information and developing contacts through which they could better, and more effectively, prepare for new funding initiatives. In contrast, cities in the southern cohesion states had traditionally much greater reliance on the national level to filter information, channel funding, and allow local interests to become involved. Inefficient national administrations, therefore, contrast with the more locally based entrepreneurial attitudes developed in the north. Additionally, the southern cities had less experience in involving local interest groups in the form of partnerships favoured by the Commission when deciding on funding, and in putting together the co-financing needed by the Programme. The URBAN initiative, in the specific context of urban policy, thus exposed some familiar problems, as well as the limits of the Commission which, however conscious of the importance of urban economies in meeting European objectives, still had to overcome fundamental spatial and national administrative differences, including inertia in changing the policy focus. The ideal city imagined in Brussels was very much a northern city and one that had shrugged off its industrial past. Southern cities, lacking financial autonomy and the technical administrative capacities of many northern cities, lag behind this ideal of northern urban entrepreneurialism.

Between 1994 and 1999, URBAN provided funds to projects in 130 towns and cities and 70 are targeted in the current 2000–2006 programme. Experiences with the URBAN programme during its first period of operation fed into the review of EU programmes for the 2000–2006 period. The partnership principle underpinning URBAN was introduced to the main funding programmes of spatial policies. Thus, the new Objective 2 of ERDF was given an explicit urban orientation in acknowledgement of the crucial economic development role of cities. The capacity of all European cities to work in these ways remains, however, in doubt, as institutional and practical legacies in policy making predispose some cities more than others to engage in *partnerships* with other scales of governance.

During the 1990s, various forms of informal cooperation flourished, complementing or substituting formal arrangements. Of particular significance has been growing informal cooperation and Commission interventions in the field of urban, as distinct from regional, policy. One such collaborative programme was the three year Urban Exchange Initiative which pulled together best practice in regeneration around a number of themes including town centre community involvement and transport policy (see, www.regeneration.detr.gov.uk/uei). While cities and some national governments may share interests in urban problems, the involvement of the Commission in urban affairs raises important constitutional issues. At the heart of this debate is the issue of the competence of the Commission. Many member states are reluctant to provoke sceptical electorates by proposing such new competencies. At the end of the 1990s the Commission itself was beset by scandal and lacked widespread public support, and the idea of a 'democratic deficit' in European institutions also worked against proposals for expanding the Commission's role in urban affairs. In contrast, leading European cities in the Eurocities network wanted the European Commission to have a competence in urban as well as regional issues (Eurocities, 1998).

The European Commission therefore stepped carefully into the urban arena and in its initial policy report pointed out that no additional powers at European level were needed but that enhanced cooperation and coordination could help the member states share experience and facilitate solutions. This document was used to begin a debate on urban issues with the Commission publishing a subsequent expanded version in 1998 (EC, 1997, 1998). The central argument of these reports is that urban areas are the dynamos of the European economy and therefore crucial in the drive to greater competitiveness. But cities are also acknowledged sites of social exclusion and contrasts. Thus, the themes of the debate are about strengthening economic prosperity and social inclusion and integrating problem areas into the wider city and regional fabric. The leading (city) regions are regarded as models for the less successful and less developed regions to follow. This venture into urban policy was encouraged by city governments to obtain some of the funds traditionally reserved for the rural peripheral areas. Most of the big cities had, by now, direct representation in Brussels to increase their political presence and lobby for funding. The leading cities of Europe are members of the 90 cities strong Eurocities network which lobbies on behalf of a self-defined set of dynamic cities and which shares technical knowledge and expertise. Each city seeks an international reputation and most have sought to enhance their international status through undertaking large projects (Gachelin, 1998). This activity is accompanied by a new spatial image of Europe consisting of pro-active, network oriented competitive regions rather than static regional 'containers' of welfare state policies (Voelzkow, 2000). The lobbying organization, Eurocities, promotes the idea of the 'functional urban area' as a basic economic and policy unit, because this enhances their political standing and influence on EU structural development policies in general, and urban development in particular (Eurocities, 1998).

The shift from regional to urban scale reflects both ever closer linkages between cities and European institutions and a new set of ideas about the dynamic properties of city regions in the European economy. The Europe of the Regions which had seemed to represent the future of European economy and institutions at the end of the 1980s has been ousted by an image of city regions as the real economic motors for, and new foci of, European policy intervention. Assisting this transformation in thinking about European space has been a radical re-evaluation of regional policy, which has gained in urgency from the proposed expansion of the European Union into central Europe and the financial affordability of conventional region-based structural policies.

Agenda 2000: streamlining and refocusing EU structural policy

The wide range and cost of regional expenditures led the EU to propose a wholesale review of its programmes in the mid-1990s. By this time the Europe of 15 member states was also considering a long list of potential new members from central, eastern and southern Europe. This combination of factors provoked a fundamental review of the core regional funding arrangements. The last systematic reform, in 1988, identified five priority objectives including support for both the 'lagging' regions, i.e. Objective 1, and for regions affected by unemployment (and numerous other factors) as a result of economic restructuring, i.e. Objective 2. The map of these areas showed that cities and regions in over half of Europe qualified under one

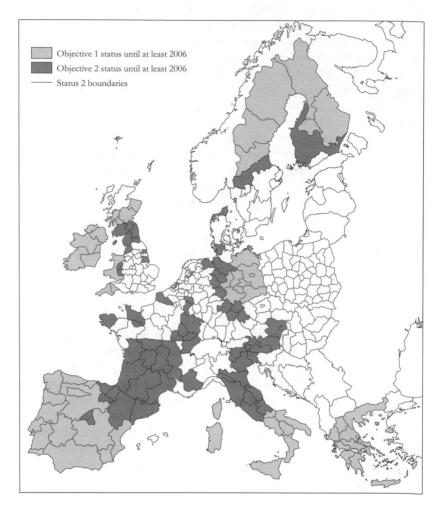

Map 3.4 Established Objective 1 and 2 status until 2006.

of the five categories (see Map 3.4), with a new sixth Objective being introduced to include assistance to the remote areas of Sweden and Finland. Such
global assistance was certainly no longer targeting the most difficult problems or regions where economic convergence was having the most serious
impacts. Additionally, it was clear that cities and regions in the countries
about to join the EU were, for the most part, worse off than those in the
existing lagging regions. Continuing with the same funding programmes
would have required a huge increase in the European budget, which would
have been politically untenable. Already, the budget for regional funding
has grown from 44 billion ecu for the 1989–1993 programme period, to
141 billion for 1994–1999 and 181 billion for the 2000–2006 period. Based

on existing criteria, regional development support for Hungary alone is estimated at between 1.5 and 2 billion ecu (Horváth, 1999). Alternatives were put forward by the Commission in Agenda 2000, published in 1997.

Agenda 2000 proposed a reduction in Objective areas and a raising of thresholds for qualification for support. The GDP of the 'cohesion' countries (Spain, Ireland, Greece and Portugal) was forecast to be at 78 per cent of the EU average in 1999 rising from 65 per cent ten years earlier. In the case of Greece, whereas in 1993 GDP stood at 64 per cent of the EU level, by 2006 this is expected to have risen to 80 per cent (Tsoulouvis, 1999). The claims for funding support from the central European accession countries waiting to join the EU will be significantly greater.

Three priority areas were proposed and built into the funding programme for 2000–2006. Regions earmarked for Objective 1 status may not exceed a GDP of 75 per cent of the Community average in order to qualify for funding. These areas are largely confined to the periphery of Sweden, Finland, Greece, the new *Länder* of eastern Germany, southern Italy, the western regions of Spain, parts of Portugal and the west of Ireland and Wales. Other areas such as the Scottish Highlands and former mining areas of Belgium, which had been supported under the old arrangements, receive some transitional support. The only urban regions in northern Europe receiving assistance are Merseyside and South Yorkshire (including Sheffield) and some smaller cities in former East Germany. The largest shares of regional funding go to Spain (23 per cent), Italy (16 per cent) and Germany (15 per cent). The new, more restrictive qualification to 75 per cent of average GDP sought to exclude a large proportion of regions that had been eligible for funding during the 1990s. For the same reasons, the revised Objective 2 status was similarly selective in restricting eligible areas to 18 per cent of national populations. In addition to support to regions in the member states, the regional budget made substantial 'pre-accession' funds available to Bulgaria, the Czech Republic, Estonia, Hungary, Lithuania, Latvia, Poland, Romania, Slovenia and Slovakia. About 40 billion ecu were earmarked for the new member states up to 2006.

It also became clear in the debate about Agenda 2000 and in the formulation of the 2000–2006 scheme that at the end of the programming period many of those regions which currently enjoy financial assistance would no longer do so after 2006 when new states had joined the EU. The existing member states agreed 'transitional' funding for some regions in preparation for the eventual loss of regional funds. For example, for the 2000–2006 period, few areas in the UK remained eligible for EU assistance. South Yorkshire, West Wales and the valleys, Cornwall, Merseyside and Northern Ireland alone retained their European funding status. Supported areas in France were reduced to Corsica and some targeted arrondissements in Nord-Pas de Calais. But the transition in regional funding remains politically controversial, with the EC hoping to defer decisions on future regional assistance for as long as possible.

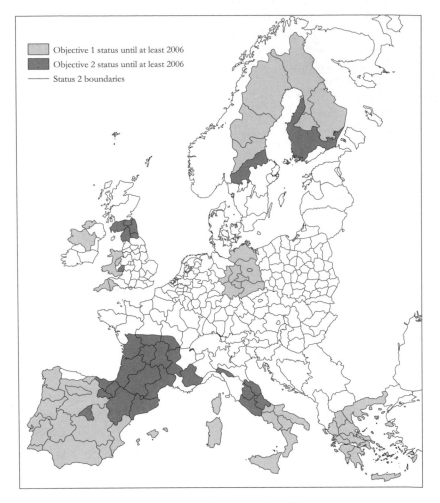

Map 3.5 Designated Objective 1 and 2 status after 2006.

The map of regional policy support has been substantially revised (see Map 3.5). As the almost blanket coverage of regional funding was being withdrawn it is not so surprising that the networks of city regions began to lobby for new European competencies and specific urban programmes or that the Commission should want to continue to assist cities and regions in some other ways. In the course of thirty or more years of regional policy the emphasis has changed from the lagging rural and outworn industrial landscape of western Europe to a new focus on employment and social issues in the major urban areas.

The European Commission provides substantial policy guidance to assist regions in the development of projects for funding. Current guidance states

that new programmes must help continue the reduction of regional disparities and establish the conditions that will assure convergence and long-term
sustainable development. This seeks to highlight the political ideal of
integration underpinning the original idea of the EU in contrast to the
continuous primary concern of member states with securing funding from
the EU. The programmes aim to achieve sustainable growth through
improved competitiveness to foster and maintain gains made in earlier
periods. For example, in relation to the programme for London, there is a
long list of policy areas to consider. These are:

- creating the basic conditions for regional competitiveness;
- competitive enterprises for employment creation;
- the European Employment Strategy;
- horizontal principles: environment and sustainable development and
  equal opportunities;
- sustainable urban development.

Cities and regions have to structure bids for funding around several
layers of European policy. The current EU Guidance makes it clear that
improvements to infrastructure are a central part of increasing competitiveness and it encourages cities and regions to bring forward projects relat-
ing to transport infrastructure, energy, telecommunications and research
and development.

The process of redefining regional problems and policies accelerated in
the 1990s. The Commission aims to influence urban policy by ensuring that
the themes of its policy statements – increasing prosperity in towns and
cities, promoting social inclusion, protecting the urban environment and
promoting urban governance – are developed through the allocation of
regional funds and through the medium of the ESDP applied to national
and regional plans. The important shift in thinking 'that occurred at the end
of the 1990s' was that dynamic urban areas can aid regional development
and European competitiveness. This is a shift of emphasis from immediate
support for the lagging and less successful to an acceptance that successful city regions hold lessons for all, and will also benefit, eventually, more
peripheral areas. The propositions of the new regionalism have a clear affinity
with the changing direction of thinking in the European Commission.
Balanced development across Europe is to be achieved by emulating the
success stories of leading cities and regions, which were largely shaped by
'good', innovative practices in governance. Following our earlier discussion
of the new regionalism we would not expect the simple transfer of lessons
to meet with much success. As we have argued in relation to the ESDP
such policy ambitions may run up against political realities and against the
dominance of market forces which tend to concentrate development in a
liberal market regime.

New institutions for European competitiveness

Over thirty or more years, European regional policies have become more sophisticated and targeted on specific regional, and more recently, increasingly also urban, problems. It would be wrong, however, to assume that European policies have been considered and applied in the same way in each of the member states. Regional funding has been channelled to regions by a range of national level institutions which look and work differently in each state. In the UK, two central government departments, the DTI and DTLR, are involved at European level. In France, national government also manages the European regional budget, while in Germany the *Länder* as regional states take an increasingly greater role in implementing EU regional policy at the expense of the national government's influence. While we can consider that urban and regional policy has been 'Europeanized', this has worked in different ways in not only different states, but also, as has been recognized, in different cities. Some have been enthusiastic supporters of a European level, setting up lobbying offices in Brussels and networking with EC officials. Others have failed to take up the challenge, relying on their national governments to hand down policy paradigms and funding.

As shown in Chapter 2, there are various theoretical accounts of how we should understand the interactions of European, national, regional and local institutions. Several institutional developments are worth examining, because they demonstrate the paradigmatic shifts in EU regional policy. At the start of the 1990s, perhaps the most significant development, or rather the innovation that was heralded as significant, was the setting up of the Committee of the Regions (COR). It was the establishment of COR and thus the formal recognition of a more visible role of the regional scale in the formulation of EU spatial policies, that led Marks (1993) to envisage multilevel governance in Europe. This local and regional government consultative committee played a formal role in commenting on projects of the Commission, the Parliament and on other relevant issues. It can be seen in its origins as one of the ways in which the Commission hoped to ally itself with subnational governments and bypass those nation states that were less than sympathetic to European intervention at regional level. The actual impact of the Committee has been limited, although it has attempted to voice opinions beyond the limited role originally given to it (Warleigh, 1997) and it continues to promote treaty revisions to enhance its status. Perhaps of greater importance over the past ten years, however, has been the influence of city networks, such as the Eurocities lobby referred to earlier. The relative greater influence of cities than the regions may be seen as a sign of the urban shift in attention in EU structural policy over the last ten or so years. The Commission encouraged city and regional networking and through regional funding programmes explicitly encouraged new types of institutions and forms of governance to emerge in the cities and regions receiving support, so that greater policy efficacy could be achieved.

A second important institutional development has been the encouragement, through European funding, of cross-sector partnership in the allocation of funding. The institutional dimension has become an explicit objective of European policy. The Commissioner for regional policy, Michel Barnier from the Savoie region, stresses the importance of developing the 'innovative capacities' of local actors involved in European funded projects (*Inforegio News*, 68, 1999). The reward for putting together the 'right', that is innovative and expectedly effective, institutions and methods of governance, for instance, inter-urban alliances, is obtaining funding. Although, as Danson *et al.* (1999) point out in relation to new partnerships in Scottish cities, alliances might only last as long as necessary to draw down European money and might not herald fundamentally new forms of policy making. Networked 'new' governance may thus be merely temporary, but it may facilitate mobilizing new capacities which may help cities and regions function without European support in the future. For the present, however, new institutions bind key actors into European ways of managing development. The major cities that have benefited from European funding over the past thirty or so years are willing supporters of such 'innovative' institutional approaches to urban and regional policy. The large cities, beneficiaries of European funding in the past, support the Commission's new interest in urban policy, and it was Birmingham, a major beneficiary of city targeted funding, which hosted the year 2000 debate on sustainable urban development. There has been an increasing penetration of European issues into urban politics. Formal partnerships may be weak and budgets limited, but cities such as Birmingham have become consciously European in outlook, not least because the change in EU policies has allowed it to step out of the confines of a firmly centralized national policy-making and funding framework. This may have important implications for the future development of city and regional policy.

But it would be wrong to see institutional development as only involving the Commission and subnational governments. In Europe's unitary states, the setting of regional budgets and selection of projects for funding continue to be tasks if not undertaken directly by civil servants at the national level, then at the least closely supervised by them. In the case of Britain, the establishment of Government Offices in the regions in the early 1990s can be seen as an attempt by national government to underscore its regional presence and retain control over the regional funding process. In the 1980s, the UK government had been one of those national governments accused of using European funds in place of, rather than in addition to, national funding. This issue came to a head over the use of the RECHAR funds in former coalfield areas and the UK government's reduction of national support to these very areas (Martin, 1999). While that particular source of conflict between the Commission and member states has passed, national governments still control the process. In Greece, for example, European regional funding continues to be entirely a matter for national government

(Georgiou, 1999). Michel Barnier's vision of local capacity building and greater bottom-up input into regional policies, therefore, presents only part of the institutional reality of European regional and urban policy. For most cities and regions the experience is of a top-down – by Commission and nation state – driven process. As programmes have developed, so funding rules and regulations have increased, because new guidelines are being established on top of existing arrangements. The resulting 'regulative thickness' and its complexity of funding programmes are long-standing complaints by subnational governments. For example, the document that underpinned the funding programme for Objective 1 in Merseyside for the period 2000–2006 stretched to 350 pages of text produced under the guidance of central government, and to be studied and vetted by the Commission. Given the bureaucratic weight attached to regional programmes, it is not surprising to find another long-standing complaint, that there is too little opportunity for local community involvement in the process accompanying the current round of funding.

Efficacy, control and legitimacy of regional policies

Another continuing complaint about European regional policy and funding is that too little is known about its actual impacts, which brings us back to the main issue examined in the early part of this chapter – the continuing inter- and intraregional disparities in Europe. The limited knowledge about impacts remains a fundamental weakness of a series of programmes which, for some thirty years, have at their core aimed to resolve issues of regional divergence and social exclusion. There are several reasons for this lack of knowledge about the effectiveness of policies. Effective monitoring of regional policy has proved elusive due to the wide range of schemes and criteria. But the problem is also rooted in administrative arrangements: for most of their life the Regional Funds have been monitored by the *national* governments who set the budgets and formulated and promoted the projects. Independent assessment has been the exception. Success in bureaucratic terms meant spending the allocated funds rather than assessing their impacts, a characteristic of public sector thinking. However, European bureaucracy is only part of the reason why little is known about actual impacts. European regional and urban funds apply now to 15 member states each with different constitutions, structures, administrative systems, central local government relationships and traditions of working to develop cities and regions. One set of rules and processes was unlikely to fit everyone and everywhere. Also, national variation can be expected to increase as new member states join. Critics of the European scale of policy making refer to a 'democratic deficit', meaning that the controls of the European Parliament are weak and that the involvement of subnational institutions in decisions is most often less than that of the nation states. Greater harmonization of policies and their implementation under the auspices of the EU rather than the nation states would also raise questions about democratic legitimacy. The Committee of the Regions periodically calls for clarification of EU, national, regional and local roles, but this perhaps reflects its own marginal status in European policy processes and within member states. As we shall see in Chapter 4, institutional variation is the norm between, and in many cases within, member states. Local, regional and national institutions adapt to the challenges of economic competition and European integration in a variety of ways. Understanding the impact of the European level also needs acknowledgement of the institutional shifts that have accompanied policy development, but those shifts also reflect a range of national and local constitutional, political and cultural differences. It is these differences that Chapter 4 intends to highlight. However, despite these complexities, the lesson drawn by the countries waiting to join the EU is that modelling regional policies according to the Commission's models is likely to assist the case for membership and that support for regional development is more likely to follow evidence of European best practice than any individualized process more suited to particular, national, regional or urban issues of economic transition.

4 Governing mono- and polycentric city regions in Europe

The previous chapter discussed the rapidly increasing importance of city regions in the context of EU structural policy and the changing policy foci and funding regimes, with cities emerging as vital centres of regional development. Attention centred on the competitiveness of cities and regions and produced a wide range of institutional responses at national and subnational level (see SPESP, 2001). Now, the focus shifts to examine how national and local factors interact to shape new regions at a range of scales, or fail to bring forward what are considered necessary institutional and policy adjustments. This includes in particular the potential implications of the forms of state organization between federalism and centralism and thus the relative degree of power devolution to subnational government. This, in turn, is likely to affect the shaping of city-regional governance. The other main factor is the underlying relationship between cities and 'their' regions. This reflects, on one hand, the principal institutional and practical legacies of urban and regional governance, and the specific city-regional circumstances, on the other. They find their expression both in the relative roles of planning and strategic policy making and in the more detailed nature of relationships between cities and regions. Both circumscribe the scope and practice of regionalization. Effectively, there are two key dimensions circumscribing city-regional governance as discussed below: (1) the external determinants as the main constitutional provisions for cities and regions to develop their specific forms of governance, and (2) region-internal factors, especially the role and position of cities vis-à-vis 'their' respective regions (see Figure 4.1).

Four main scenarios emerge from overlaying the constitutional and region-structural factors as shown in Figure 4.1: the varying degree of power devolution to subnational government under federalism and centralism, combined with the main structural difference in the nature of regions between a monocentric and a polycentric arrangement. In the former, one city dominates the region while, in the latter, several cities are sharing in, or competing for, a regional role. In a *monocentric* region, one dominant metropolitan core may be reducing the region to effectively little more than its own hinterland, while in the case of *polycentric* regions at least two

		External factor: devolution: unitary versus federal state organization	
		Unitary state structure	**Federal state structure**
Intraregional factor: mono- or polycentric structure	**Monocentric**	**Scenario 1** Single city with limited autonomy and detailed state control	**Scenario 2** Single city with considerable policy-making autonomy
	Polycentric	**Scenario 3** Competing cities with limited autonomy and detailed state control	**Scenario 4** Competing cities with considerable policy-making autonomy

Figure 4.1 Principal intra- and extraregional institutional determinants of city-regional governance: structure and constitution.

competing urban centres give the region a theoretically greater influence vis-à-vis the localist competition. The inherent inter-urban competition may limit any willingness to cooperate, if there are perceived local costs in financial or political terms. Any identified common policy objective will sit opposite the forces of competitive localism. Economic necessities, such as competing for limited volumes of public or private investment capital, for instance, may encourage non-cooperation, while the growing complexity of development issues, and the growing emphasis on the regional scale in corporate investment decisions, push for more regional engagement by policy makers. Polycentric regions seem inherently more affected by these essentially contradictory aims of city-regional governance than monocentric regions, where the dominant city's interests and policies are likely to shape the agenda for the rest of the region's governance. As a result, a clear regional identity and 'voice' is more likely to emerge in monocentric than polycentric regions with their latent internal divisions (see also Stiens, 2000). It is these differences that underpin the following discussion of country-specific examples of urban-regional planning and policy making (see Table 4.1), looking in particular at a range of institutional and policy responses to the changing context of a competitive Europe.

There is certainly a wide variation in the roles of the different government tiers in managing city-regional responses to competition and to European policies. Table 4.1 summarizes the institutional, constitutional and scalar characteristics of the regional examples discussed below, which largely circumscribe regionalization (see Figure 4.1). The main divisions between

Table 4.1 Mono- and polycentric regionalization across Europe

Example	State organization	Type of region	Constitutional context	Scale	Key initiatives
Finland	unitary	large subnational entities (mono- and polycentric)	EU-inspired regions, regions as collaboration of generally small Local Authorities, no regional government	LA groupings, urban hinterlands	establishing regional councils as 'add ons' to established planning hierarchy
France	unitary	'super regions' (polycentric due to scale), 'dynamic cities' (monocentric)	competitiveness-inspired regionalization of and within *départements*	groups of regions (immediately subnational), groups of urban communities (intercommunal level)	new urban regions as growth centres, regional councils (marketing and public transport/tourism)
Netherlands	unitary	polycentric: as national subdivisions and agglomerations (Randstad)	weak regional administration, central control of finance, formalized planning hierarchy competes with informal regional policy arrangements	subnational country-wide subdivisions, metropolitan regions	establishing regions across country for planning, effectively powerless 'containers' for policies, urban regions as collaboration between neighbouring large cities
Italy	federal	provinces (polycentric) and metropolitan regions (monocentric)	traditional city-focused regions (spheres of influence), provinces as historic spaces, limited powers, state controlled resources	provinces as formal regions country wide, metropolitan regions as functional hinterlands of main cities	raising profile of provinces for economic development and tourism marketing (place marketing), encourage less formal regionalism (city-based)

Table 4.1 continued

Example	State organization	Type of region	Constitutional context	Scale	Key initiatives
Spain	federal	autonomous communities (ACs) (polycentric) and urban regions (monocentric)	devolution of cultural provinces (Andalucia), strong local (urban) govt.	immediately subnational large ACs, city regions as urban hinterlands	city-based regionalization (e.g. Barcelona) for internal competition, city networks, EU and state sponsored city regions
Germany (see also Chapters 6, 7)	federal	'provinces' (Länder) monocentric (city states) and polycentric (see Chs 6, 7)	devolved federal state, strong regions and LAs, elaborated planning hierarchy	Länder as main actors in regionalization, policy-based collaboration between LAs	required regional plans and development strategies, 'bottom-up' regionalization with more flexibility in range of actors
United Kingdom (see also Chapters 6, 7)	unitary	England, Wales, Scotland, N. Ireland as parts of UK (polycentric), new regions in England (mono- and polycentric)	newly empowered components of UK (devolution), new regions in England as marketing and policy regions	devolved nation-regions, English regions as 2nd tier, counties and informal interlocal collaboration below that	devolution of UK, in England, raised profile for old planning regions, business focused regional development agencies as main actors, little democratic representation

the cases are the external state-constitutional arrangements stretching between federal and unitary, and the internal structures of the regions between monocentric and polycentric arrangements. Scales vary between large subdivisions of nation states on one hand, and smaller city-regional areas. The table also shows the diverse nature of initiatives and policy measures that emerged from national and regional considerations. This includes the need to address and overcome divisive localism in the interest of genuinely regional initiatives, for which a clearly defined common objective, involvement of civil society and sound funding of regional institutions are deemed essential (SPESP, 2001). Considering constitutional factors among the external determinants of regionalization, German city states represent one end of an institutional spectrum of city-regional power, and the indirectly elected and financially dependent Lisbon Metropolitan Authority, the other.

The differences among the case studies allow us to distinguish three main types of regional planning systems, with varying degrees of formalization and regional power vis-à-vis the local authorities: at the one end, there are loose, informal arrangements with clearly circumscribed and limited powers, guarded jealously by the relevant localities (see e.g. the Local Authority Association Ruhr), at the other there are strong regions acting top-down as agents of central (national) government and affecting local planning directly (e.g. Joint Planning Body in Berlin-Brandenburg or Regional Council of Île-de-France, Paris). Sandwiched in between the two are spatial planning systems with locally based, bottom-up forms of regionalizaton within a more or less centralized state (Lutzky *et al.*, 1999) (e.g. Metropolitan Agency in Lisbon or cooperation in the Copenhagen Region). The voluntary arrangements are a clear indication of the established territorial planning structures and responsibilities being insufficiently congruent with existing functional territories and related policy requirements (policy territoriality). New self-help solutions are thus deemed necessary in the shape of informal, single-purpose collaboration. The outcome is loosely integrated and limited life 'designer regions' (Weichhart, 2000). The Berlin-Brandenburg and Yorkshire and the Humber regions provide such examples, as discussed in the next chapters. Examples of policy driven alliances and narrowly targeted (if time limited) tailor-made regionalization include concerns with green belts, such as Berlin's eight regional parks surrounding the city, the Green Belt plan 1995 in the Paris-Île-de-France region or the regional agricultural park 'Milan South' (Lutzky *et al.*, 1999). Other examples of purpose-driven 'designer regions' include waste management (e.g. Helsinki and Lisbon regions) and public transport.

Constitutional differences alone, however, apart from the very nature of the region as a constitutionally established or temporary arrangement, do not explain the differences in regionalization. National traditions in government, party political factors, cultural differences and historic-political legacies interact with the formal structures of government to push and pull cities and regions towards more effective governance. Differences in practical regionalization will thus be shaped by more than merely administrative differences between federal and unitary states, and the number of key cities in a region in different parts of Europe. The cases examined here point to the significance of the quality of practical interactions between national, regional and local institutions, and support many of the arguments examined in Chapter 2. The cases also show the relevance of the nature of city regions between mono- or polycentricity, and the fact that city-regional relationships can be shaped through administrative adjustments as part of technocratic solutions to urban and regional problems by the governments. Alternatively, the new governance may bring forward development coalitions of cross-sector interests within and between territorial scales, that can push institutional and policy reform. The latter approach may appear more acceptable under localist competitiveness, as it permits an entirely pragmatic issue-based and time-limited collaboration with little effective commitment.

Constitutional changes may seem the obvious and immediately effective strategy to further regionalization. Such fundamental changes to existing power distribution face, however, considerable resistance from local authorities wary of losing policy-making autonomy. Where changes have been implemented, mainly driven by inherent strong cultural-historic regional identities, constitutional reform has had a profound impact on the role of regions, such as in the UK, in Belgium and Spain, where the changes were part of modifications to the nation states' structure. Thus, Belgium and Spain embarked on devolution to accommodate different cultural and linguistic subdivisions across the nation states. In the Belgian case, the state was transformed from a unitary into a federal structure with now three autonomous, cultural regions (roughly north and south of Brussels, with Brussels a separate entity). In Spain, inherent regional cultural differences and pressures towards greater regional autonomy have also led to federal structures, with considerable autonomy, especially for Catalonia. The French government also experimented with a very limited form of devolution to Corsica to accommodate strong regionalist, even separatist, tendencies, and in Italy several 'special regions' were established to acknowledge an underlying greater sense of historic-cultural identity, e.g. Sardinia or Tirol. The measures finally agreed in summer 2000 offer only temporary transfer of some powers after 2004, reflecting the bargaining between regional separateness and national control. In the UK, limited devolution to Scotland and Wales was introduced at the end of the 1990s. This reform responds to nationalist demands and to the strong influence of the Scottish and Welsh Labour Parties. However, relationships between central government and the new national parliaments have not been smooth, with the former trying to minimize its loss of power. The limited devolution was initially seen as an overture to possibly much further-reaching devolution, particularly in England, but such possibilities now seem less likely to materialize. Only in London has a democratically elected regional representation been established. Such city region based changes to governance will be discussed for the Netherlands and Italy for polycentric regions, and for Germany, exemplified by the Rhine-Ruhr region, in the following chapter. Approaches to the city-regional governance of monocentric regions include those in Portugal, Denmark, Sweden and also Germany.

The external constitutional factors will circumscribe not only the willingness and scope in principle for collaboration between cities within and between regions but also the very nature of any collaboration. The main differences involve the 'formality' of the arena of cooperation, and the public or private sector nature of actors, i.e. the emphasis on government or governance. On that basis, four main forms of cooperation may be identified across Europe (SPESP, 2001) as a combination of public and private sector actors on the one hand (emphasis on government or governance respectively), and formal or informal platforms of communication on the other. The *communautés de ville* in France are typical examples of a solely public sector actor-based arrangement (i.e. 'government') on a formal platform.

While, as we saw in the arguments reviewed in Chapter 2, new institutions and policy at city-regional level may be desirable, as yet there are few cases – Stockholm, Vienna, Hamburg and Madrid may be the best examples – where urban functional areas coincide with administrative boundaries. In the case of Stockholm city region, Stockholm-Mälar, the urban core consists of two urban authorities, and there are some 40 other municipalities and four intermunicipal collaborative associations sharing in the region's governance. Established in 1992, the region is an informal organization offering a framework for collaboration between the various actors in the region. This arrangement has no fixed territoriality as it is essentially a imagined space, but it is defined by the territorial sum of the participating municipalities' areas. The notion of the region's territory is thus inherently variable (SPESP, 2001). Even in such cases where economic and institutional boundaries fit, actual competence at this level may be limited. In Helsinki, for example, the functions of the Metropolitan Area Council are essentially limited to waste management and public transport. At city-regional level it is often the overlapping jurisdictions of local and national governments which affect the development and implementation of regional policy. In federal Austria, for example, national-level coordination is attempted through an informal platform for discussion, the Spatial Development Conference for Austria (*Österreichische Raumordnungskonferenz*), in which all levels of government are represented but whose prescriptions are not binding. As in Germany, it is the federal states (*Länder*) which retain the main competence in regional planning. At city-regional level problems of coordination are addressed through informal collaborative arrangements, such as planning agreements between neighbouring municipalities. One example, in the Vienna region, is the Planning Association-East (*Planungs-gemeinschaft-Ost*) between the city of Vienna and the neighbouring *Länder* of Lower Austria and Burgenland.

In Denmark, for instance, the new national Planning Report (DMEE, 2000) emphasizes a shift away from institutionalized planning to a more policy oriented, pro-active way to facilitate economic development. This includes stronger collaboration between government levels and between governmental and other actors. While there continues to be a strong emphasis on conventional, territorial-based regional planning, the new over-arching agenda across Europe is on network building between local authorities, cities in particular, and the inclusion of regional considerations in local planning (DMEE, 2000; Stiens, 2000). The new emphasis on 'poly-centric city systems' (DMEE, 2000: 16) includes the designation of 'national centres' as cases of good practice in regional network building both between cities and between cities and rural areas. 'The designation of national centres recognizes that some cities and also city networks must serve as locomotives in regional development. The national centres therefore contribute to recog-nizing the mutual dependence between town and country' (DMEE, 2000: 19). This specifically urban focus in national regionalization finds some echo in the new German emphasis on seven European Metropolitan Regions, which were introduced to the national planning paradigm in the late 1990s (Stiens, 2000), although there, the emphasis is more clearly on international competitiveness and networking.

Increasing European cooperation and networking, of course, has had an impact on urban areas. In practical terms, however, it has been more through traditional territory-based regional policies, i.e. Objective 1 and 2 designa-tion, than the more recent city focused URBAN programme, that new 'partnerships' have been encouraged. In some cases, the evolution of the Structural Funds has had a direct impact on regional boundaries which are being drawn to maximize the chances of obtaining grant aid. For example, the economically successful Dublin region was separated from the rest of Ireland in order to safeguard Objective 1 status for other parts of the country. In other cases, however, historic administrative factors are the main reasons behind such divisions. In Denmark, the Copenhagen metropolitan region is divided between the twin core cities and unitary authorities of Copenhagen and Frederiksborg, and the surrounding Copenhagen County with its own administration (see Map 4.1). To cover the whole functional region, two more outlying counties would need to be included. In some respects, the situation is thus not unlike those in London and Berlin, where core city and hinterland region are separated by administrative boundaries, as discussed below in more detail (see Chapter 6). Indeed, the divisions have been a result of following London's example and abolishing Copenhagen Council in 1990. The reasons were similar – central government disliked the strong political power amassing in the capital region with about one third of the

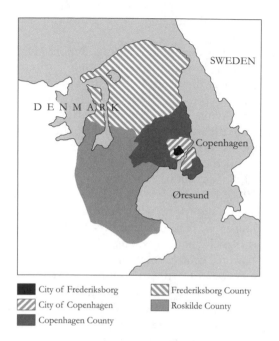

City of Frederiksborg   Frederiksborg County
City of Copenhagen   Roskilde County
Copenhagen County

Map 4.1 The Copenhagen region.

population (interviews with Copenhagen County and Greater Copenhagen Authority, 16 Jan. 2002). Region-wide planning was left, until last year, to the five local governments sharing into the metropolitan area. The similarity with London continues with the establishment of a Greater Copenhagen Authority in 2001 as a strategic body responsible for transport and regional planning, but with few real powers. Politics were the main reason for choosing the 'weak' solution for regional governance. The two mayors of the largest municipalities, Copenhagen City and Copenhagen County could not agree who should take the lead in the unified structure (interviews at CC and GCA, 16 Jan. 2002). Achieving an agreed, common objective requires some common interest as a rallying point for joint policies. Otherwise, the core cities may pursue their own interests and bypass the surrounding region. In Copenhagen's case, for instance, the City Council takes an explicitly outward looking, international perspective, thus seeking to boost its competitive visibility. This is indicated both by its own representation at the European Commission through its liaison office in Brussels and its active participation in the Øresund economic region. This is a city-regional network established between Swedish and Danish local and regional authorities around this part of the Baltic coast. The wider Copenhagen city region does not directly participate in this network, but benefits through the core city's links. One indirect result of the competition with Sweden in the Øresund region was the extra support it generated for establishing some, albeit weak, regional structure in the Copenhagen area. Sweden has just established a new regional structure for the Scania region, and this created an institutional asymmetry between the two neighbours and made communication at subnational level more difficult.

These city-based initiatives illustrate an observed general shift across Europe from traditional Keynesian-style, welfare oriented, redistributive, territorial policies based on large centrally managed financial transfers between regions, towards a more city focused, network-based, individualized, and inherently competitive planning for activating growth (see e.g. Voelzkow, 2000).

Constitutional reform, party political projects, global competitiveness, European grant chasing and emphasis on 'harmonization', all help reshape urban and regional governance across western Europe. This has, since the 1980s, included various attempts at creating new institutions and policies under the varying constitutional and geographic-structural circumstances found in Europe. Seeking to attract prestige international events has been one strategy. Glasgow and Copenhagen, for instance, benefited from designation as European City of Culture, while Liverpool hopes that its bid in this competition will help change its image. Lisbon benefited from substantial national infrastructure investment to support Expo 98. Barcelona, which has become somewhat of a cause célèbre in city-regional discussions, used the Olympic Games to drive forward urban and regional restructuring, under the leadership of its mayor, and advanced its leadership in the region. There are no guaranteed benefits, however, as the examples of Seville and Hanover have shown; both gained rather less from hosting Expos, largely due to less than ideal image creation. Even failed attempts, such as Manchester's unsuccessful bid to attract the Olympics, can, however, generate some positive effects, such as facilitating institutional and policy reforms and new formal and informal contacts, which can be used as foundations for bids for subsequent projects of economic revival.

At a larger regional scale, the centralized French state created a tier of elected regional government in the 1980s, but the regional problem in its various forms, such as uneven development and urban–rural contrasts, has not been resolved. Nevertheless, the French experience offered the UK government a set of experiences to learn from, although there are marked contrasts between the French approach and that of the resurgent English regionalism. There have also been more fundamental constitutional shifts to federalism, such as in Belgium with the separation of the state territory into three autonomous regions in response to cultural-historic divisions. This chapter contrasts the relatively new problems of Belgium's regions with the complex multi-tiered and competitive institutionalism in an established federal system such as Germany. This encompasses several territorial scales between the autonomous local level and the provincial (*Land*) scale of governance, and is exemplified below by the polycentric Rhine-Main (Frankfurt) region and the monocentric Stuttgart area. City-regional issues in a quasi federal state setting are explored *inter alia* in Barcelona, while those in a centrally organized state structure are examined for the monocentric case of Lisbon or the polycentric city region of Randstad in the Netherlands. Here, regional ambitions interact with a strong national planning tradition leading to new forms of cooperation. The end of the chapter discusses the different ways in which the local, regional and national levels interact with European and competitive forces.

National constitutional contexts are of fundamental importance for shaping the ways in which regions and regional policies can work. The following sections look at the two main contrasting constitutional groupings of countries around federal and unitary arrangements respectively. The underlying, quite different, provisions for regional policy making will be discussed in terms of their impact on policy responses to the particular challenges of city regions and their governance. The first part examines federal, the second part unitary state structures and their respective impact on regionalization in different countries across Europe. By their nature, the examples represent circumstances confined to particular national territories. There are, however, developments towards transborder regionalization, encouraged by EU policies. Increasingly, these seek to facilitate cross-border regional collaboration between local and regional actors to overcome the divisions of national boundaries and allow more effective policy responses to emerging or latently underlying functional regional dynamics. This now also includes central European countries along the former Iron Curtain in preparation of the EU's eastward expansion.

These initiatives include traditional territorial regionalization, as illustrated by the Euregio Egrensis along the German–Czech border, but also network-based initiatives with lesser territorial visibility, such as the Union of Baltic Cities. In the former case, EU funding seeks to facilitate regional collaboration across the former Iron Curtain. The outcome may be the re-emergence, if tentatively, of old regional identities and affinities, but also new competitiveness among neighbouring regions and their cities. In the latter case, key cities and their international competitiveness are the main driving force of collaboration across a region, drawing on shared economic, cultural and geographic features. These two main approaches to cross-border regionalization are discussed in the third sub-section of this chapter. Although most such cross-border regions will, in effect, be a special version of the polycentric (peripheral) variety, monocentric regionalization may emerge in those cases with an asymmetric degree of urbanization on both sides of the border and a considerably larger city on one side than on the other.

Mono- and polycentric regions in Europe's unitary states

The first group of examples comprises city regions within unitary states. The strong role of central government is the key determining external factor in regional policy making and planning. There is an inherent, distinct top-down approach towards regions with the central state taking a direct role in regional governance either in place of a separate regional government altogether, such as in England, or through a somewhat less hands-on approach by devolving some responsibilities to tailor-made regional institutions under the auspices of the state. The latter case reflects the recognition that the challenges of international economic competitiveness can be better met by enhanced direct regional input into policy responses, particularly in large metropolitan areas. Finland is one such example. The examples show the internal differences between monocentric and polycentric structures and their varying ways of identifying and operationalizing regional interests.

Central–local duality in urban regions and regional planning in Finland

Finnish regional governance, rooted in Scandinavian tradition, is based on a competitive dualism between strong local autonomy as part of municipal self-government and direct central government engagement through strict guidance, particularly in local government's role in welfare provision, education and health care. It is in these areas that local authorities team up to jointly provide the statutorily required services and thus benefit from economies of scale. In 1997, there were 67 urban municipalities (out of 450 local authorities (AFLRA, 2001)) which provided the required minimum range of urban administrative functions (Antikainen, 2000). Continued rapid urbanization gives cities an increasingly prominent role in Finnish spatial development and policy at the expense of traditional redistributive, centrally managed regional policies. This goes in tandem with a support for greater local (urban) policy-making autonomy which, in turn, has also highlighted the need for greater interlocal cooperation to address regional issues (Holstila, 1999: 1).

The state, in unitary tradition, has taken a direct interest in urban and regional matters by designating 27 urban programmes as a time-limited local–central collaboration to foster local development objectives with an implicit, rather than explicit, visibly institutionalized regional perspective to it. In Helsinki, for instance, there are three such programmes concurrently active each with a narrowly defined objective in housing, city marketing and city region-wide transport. This central–local programme reflects central government's recognition that cities are the main foci of economic development across regions and that public investment needs to be channelled accordingly, albeit through specifically defined projects rather than a general policy-making empowerment at the regional scale.

These urban focused policies accompany modifications to conventional regional policy through 19 formal, state-designed regions, to allow regions to operate as vehicles for EU Structural Fund initiatives. Thus, in 1994, a new regional level of governance was introduced based on functional regionalism. Districts (*seutukunta*) were created by the state between the municipal and county levels with a distinct urban focus and a mono- or polycentric internal structure. The 88 (85 since 1997) functional districts (compared with the 19 'conventional' regions) were defined by the central state on the basis of travel-to-work areas and existing patterns of cooperation between municipalities. These districts have become an important scale for the implementation of EU regional policy (Antikainen, 2000: 9). Crucially, these newly defined functional regions transcend existing administrative boundaries (Antikainen, 2000: 8), and thus require intercommunal cooperation.

The state-designed regions are administered by regional representives of the central state and local government-based Regional Councils which receive their legitimacy indirectly through the participating local authorities. The councils' main responsibility is regional development aimed at reconciling central and local government interests, and implementing EU Structural Fund policies. Furthermore, they offer a regional focus as providers of regional services, such as research and marketing, and facilitating interlocal collaboration and communication. Steering committees seek to coordinate local interests towards a regional agenda, which is not always easy because of the established principles of local self-government and thus inherently localist tendencies. As a result, the councils may be rather fragmented in their objectives owing to the multitude of local governments involved. In principle, therefore, the Regional Councils are not dissimilar to the English Regional Chambers or the German Regional Planning Associations. By their nature, the new regions appear to be bolted on to existing arrangements, primarily to comply with EU common practice in regional policy making, rather than reflecting a genuine restructuring of regional governance. For instance, regional development plans are mostly generalized synopses of the relevant local plans, rather than genuinely *regional* plans to guide more detailed local decision making. Without a common agenda the efficacy of locally defined regional initiatives and thus the relevance of a region as a territorial scale of policy making is weakened, as highlighted by two case studies from northern and southern Finland (Sotarauta and Linnamaa, 1998). They show the importance of clearly identified common regional policy goals to act as 'rallying points' for the various local and regional actors. This has become more important with the greater ongoing suburbanization pressures which create an ever widening rift between functional and established policy-making territories (regions). Interlocal collaboration will depend on a sense of common purpose particularly in polycentric regions with competing equal cities, or, in monocentric regions, on the ability, and/or willingness, of the dominant city to lead the agenda. These two scenarios are illustrated by the southern Finnish region of Sainäpaapurit and the Helsinki region (Uusimaa region) respectively.

The Sainäpaapurit region exemplifies the importance of common structural economic characteristics and thus converging interests and policy agendas for developing a form of regional cooperation between local authorities, and a sense of shared interest. Referred to as an 'entrepreneurial region'. This suggests a distinct common denominator. Such a claim is, in practice, only appropriate for the northern half of the Sainäpaapurit region with a strong SME culture, whereas the southern half is largely agricultural. Policy objectives will thus differ between the two parts of the region, making the design of a common regional agenda difficult. With such inherent divisions and thus weakness of regional purpose, it is not surprising that the main city and economic centre of the region, Seinajoki, pursues its own agenda with greater emphasis on local qualities and network-based linkages with cities (and regions) outside its own region. As a relatively successful local economy with a thriving SME culture, the city sees little benefit in tying its fortunes to a seemingly less successful region and seeks its contacts and collaborative partners further afield. They include central government and other national organizations. The situation thus shows interesting parallels to that of Leeds in the Yorkshire and the Humber region (see Chapter 7), which, too, views the region more a liability than an asset and as something to be used merely to pursue its own interests. In such cases the externally imposed (by central government) requirement of regional cooperation tends to lead to little more than talking shops in which formulating required strategies becomes merely a ritual (Sotarauta and Liannamaa, 1998: 516), while real policies are defined and targeted elsewhere.

Scope for genuine regionalization and policy making looks better in a structurally more homogeneous region where there is a general sense among all relevant localities of participating in, and benefiting from, engaging in regional initiatives. Under such conditions, a city, which is dominant by its sheer size and/or function, can then effectively turn the rest of the region into its hinterland, because it views engaging in the region as beneficial for its own ends. Such a situation can be found in the Oulu region around the city of Oulu, the largest northern Finnish city with almost 110,000 inhabitants. Initiated and managed by the city, a number of overlapping and interrelated networks have developed both vertically across administrative scales, and horizontally within the region and beyond. The city's prominent position is reflected on the Regional Council in the fact that it sets the policy agenda (Sotarauta and Linnamaa, 1998: 518). This local focus also reflects a relative weakness in regional traditions, with much of the regional debate and arrangements merely following the requirements set out by EU policies, rather than a genuine belief in the benefits of a regional approach among local authorities (Sotarauta and Linnamaa, 1998: 517).

Helsinki is the other obvious example of a monocentric city region. The wider Helsinki region (Uusimaa region) as functional entity (commuting region) comprises 12 municipalities, of which Helsinki city council is the largest. Helsinki metropolitan area consists of four of those 12 municipalities. The city recognizes the importance of good cooperation between them, but also aims at broadening the range of regional and local actors involved from both public and private sectors in the sense of regional governance. The Helsinki Metropolitan Area consists of four municipalities, including the city of Helsinki itself. There are thus three tiers of territorial scale affecting Helsinki (at subnational level): Helsinki City, Helsinki Metropolitan Area and Helsinki Region (Uusimaa). To this end, the Helsinki mayor took the lead in shaping a city-regional agenda by establishing the Helsinki Club in 1997 as a round-table collaborative platform. This initiative shows the importance of personality in driving regionalization processes, similar to the Barcelona example. The Helsinki Club includes 17 policy makers from both the public and private sectors within the region. One of its main objectives has been to analyse the role and competitive position of the region in an international context, reflecting the international ambitions of Helsinki and the region. The role of the club has been to stimulate a political debate within the wider region and provide a communication platform for policy makers to establish contacts and mutual trust. The task then is to translate the agreements into actual initiatives, such as development plans (Holstila, 1999) and policies, which has to happen through the established statutory, mainly local, channels. The regional council's main goal is to further the capital region as Finland's main business area, and this involves collaborative consultation through the assembly delegates and aims primarily at international cooperation in the Baltic Sea region (Uusimaa Regional Council, 2001). In the case of Helsinki's region, Uusimaa, there are 72 local councillors delegated to the regional assembly, chaired by Helsinki's mayor, and it will take a clear vision for the region's future development, as well as a sense of shared success, to bring them all in line with an agreed regional policy. The experience in the Helsinki region is not unlike that of the Stockholm region in neighbouring Sweden. There, too, the capital city is by far the largest urban centre and automatically dominates the city region. Under such conditions it is crucial for fostering regional involvement by the other, smaller municipalities, that collaboration is voluntary, informal and perceived as 'safe' vis-à-vis the main city's domination in size and importance. If the initial tentative contacts can establish mutual trust, as achieved in the Stockholm-Mälar region, then the initial loose, informal arrangement could become more permanent and formalized, because the smaller authorities realize that their autonomy is not under threat from 'big brother'.

The Finnish case thus exemplifies a general shift from state-designed regions to locally based definition of contents, purpose and function of regionalization. And in this context, the importance of clear, generally agreed regional agenda and policy goals between local authorities in a region seems crucial for any genuine form of regional policy making. Otherwise, the strongest and relatively most successful city (or group of cities) of a declared region will seek to pursue its (their) own agenda without consideration of regional issues. In monocentric regions the outcome will be a region reduced to being subject to the main cities' local initiatives, and in polycentric regions, interlocal competitiveness is likely to undermine any sense of common regional purpose. The advantage of a monocentric region is the clearer leadership structure, because the largest city drives the agenda, while unequal development between the core city and the rest of the region may be a drawback.

Super regions and dynamic cities in France

There are profound tensions between the centralist French republican tradition and the notion and principles of regionalization. Nevertheless, the French government has embarked on a process of devolution of responsibilities for regional matters to the regional scale. This move has been driven by the centre's new concern with the competitive position of large urban areas, including ideas of specialist city-regional economies. This differs from the traditional view of regions as effectively being nothing more than containers for EU regional policies under the Structural Fund. These changes, however, have had only limited impact on the ground (Sallez, 1998) due, in part, to tensions between government appointed prefects as the traditional centres of regional administration, and the elected new regional and established local governments. This decentralizing shift to formal regions in the mid-1980s needs to be seen not as the emergence of an autonomous regional sphere but as a stage in the evolution of central–sub-central government relationships. Balme (1998: 182), for example, describes regional reform as a 'regionalisaton of public policy', still led by the centre, rather than a constitutional shift. One reason may be the perceived inherent weakness of subnational government. French local government operates at three spatial scales: local with 36,763 municipal authorities, 100 *départements* (including overseas), and 26 regions (of which four are overseas). The relative powerlessness and policy-making weakness is particularly apparent for the local level, where one in nine communes, i.e. 4,082, have fewer than 100 inhabitants (Négrier, 2000).

The large regions were established during the 1980s' reform, creating 22 elected regional councils with a range of responsibilities including a role in transport and economic development. They have been equipped with some independent income from taxation, but this is rather restrictive, limiting the actual competencies of regions, in addition to the competitive challenges from the other units of sub-central government.

With formalization since the 1980s, institutions and players seek to secure their 'home turfs'. One of the main problems is that of considerable institutional 'thickening'. 'French decentralisation was implemented without adapting the existing territorial jurisdictions.' Thus, the three concurrent tiers of subnational government have created, 'one of the densest networks of elected officials and bureaucrats in the world' (Négrier, 2000: 254). As a result, there is a considerable difference between constitutionally provided and practically available powers and competencies. While the region is notionally the main responsible scale for economic development, all three tiers of subnational government 'competed in inventiveness in creating their own instruments' (Négrier, 2000: 255). Not surprisingly, subsequent action 'did not run according to a logic that led to coordination, but, on the contrary, to competition and redundancies among executives' (Négrier, 2000: 255).

Such power-securing governmental competitiveness has also been found elsewhere, e.g. in Britain and Germany, where performance pressures are likely to lead to less rather than more cooperation, especially if non-cooperation yields direct political rewards. Under such circumstances, the lesser emphasis on territoriality and boundedness in French decentralization, the demarcation of legal boundaries of responsibilities and powers has become more important between competing administrative scales, because ambiguity about the limits of powers encourages expansionist ambitions between subnational governments.

The inherent difficulties with decentralization have effectively strengthened the role of the regional state within regional governance as the most immediately visible representation of regional power (Négrier, 2000). This predominance has been further helped by its role as facilitator in the implementation of EU Structural Fund policies. 'Thus, in France, the implementation of the principle of subsidiarity paradoxically often led to the empowerment of the State itself' (Négrier, 2000: 256). As a result, 'the political bankruptcy of subsidiarity thus helps justify doubts regarding its contribution to territorial redistribution, and thus a new dominant regulation discourse' (Le Galès and John, 1997; Négrier, 2000: 256). So, devolution does not automatically facilitate greater evenness in territorial competitive capacity or opportunity. This casts doubt on the essential paradigm underpinning regionalization, i.e. the shift to more governance rather than state-dominated government (see Chapter 2). It rather points to the importance of how such regionalization and devolution is actually implemented on the ground. In France, and there are some interesting parallels to the UK, the role of the regions has been defined by inadequate funding, lack of an institutional track record (experience), insufficiently developed scalar territorial and institutional identities. Many of its obvious partners operate at other territorial scales (within other boundaries), and the essential nature of the regions is effectively as think tanks developing strategies implemented at other levels, rather than being implementors themselves (Négrier, 2000: 257). Some issues have been addressed, e.g. funding and expertise, but there continue to be distinct weaknesses in the regional scale of governance. The debates around the role of the Île-de-France region around Paris as 'growth' pole and its likely future development, illustrate the importance of the centre and the changing political discourse (Lipietz, 1995).

These weaknesses include a continued detailed direct involvement by the central state. Nevertheless, there are differences. In complex intergovernmental relationships some new regions play a fuller part than others (in Nord-Pas de Calais, for example), placing a greater challenge to the centre's involvement. But central government has used its four-yearly contractual negotiation with the regions to impose national priorities rather than allow a bottom-up regionalism to emerge. The Contrat de Plan pulls together state and regional actors around development and other projects but the state has continued to dominate negotiation around the Contrat and thus the policy goals and scope for implementation.

After the 1997 national elections tensions about regionalism were further enhanced by separatist movements in Corsica and policy disagreements within the government. Thus, on the one hand, demands from Corsica for greater autonomy continued to challenge the republican tradition, leading to the resignation of the Minister of the Interior who was committed to the single, national scale of government. On the other hand, the new environment minister in 1997 personified the alternative view and attempted to shift power towards the regional level, especially around cities. The Loi d'Orientation pour l'Aménagement et de Développement Territorial, envisaged a new system of contracts from 2003, this time at the city region level. This city region scale requires cooperation among the lower level of communal governments and 142 urban areas have been identified where such intercommunal cooperation would be encouraged and local taxes harmonized. Collaborative initiatives, or outright alliances between local authorities, encompass economic development, including tourism, environmental and social and cultural projects. Three main forms of collaboration emerged, as illustrated by the case of Parthénay in the wider Poitiers area (SPESP, 2001): (1) horizontal links between the dominating central city and the other municipalities in the region, (2) cross-sectional links through public–private partnerships between business and administration, and (3) vertical links between the different tiers of government, *département*, region, state and the EU. The common agenda in this case, keeping the various actors together, is the strategic project 'digital towns' which Parthénay established with three other EU cities. Since 1995, the mayor of Parthénay has been driving the agenda towards creating a European consortium of public and private actors who are interested in the creation of an 'electronic society' with his city taking a key position in the envisaged network. The other communities of the region are tied into these plans via Parthénay as their regional centre. This highlights the importance both of a key personality to establish and drive a unifying regional agenda, and the role of cities in anchoring their regions into wider collaborative networks.

The new focus on urban regions in French regionalization politics was accompanied by a wholesale review of the competitiveness of the 22 formal regions. The national planning agency, DATAR, proposed a system of super regions, which would group together the formal regions on the basis of structural similarities and interrelationships. The result would be a new regional map of France with six super regions, including a group of regions around the Paris Basin and another in the north. In national planning and through the Contrat de Plan the state could coordinate the development of these six groupings, thus effectively continuing to have a leading role in regional matters. But this large-scale national planning view is balanced by acknowledging the dynamic role of the major urban areas in regional development across the nation state. In each super region it is the cities where economic performance has been strong which will take a 'natural' leading position in regional development policy making. The emerging national view is therefore of a state comprised of both super regions as large territorial entities with undertones of traditional state driven interventionist policies, effectively taking on some of the managerial roles of the central state, and dynamic cities with a stronger locality focused economic potential and thus scope for more state independent policy making.

Central state and new regionalization in Portugal

Portugal has a strong centralist tradition. The central state created and ran local and regional economic development programmes. Emerging from a state dictatorship, the 1976 revolution envisaged a more decentralized state with a three-tier structure of local, regional and national government. In effect, only the local tier (comprising parish and district councils) and national tier were institutionalized, without the intermediate regional level of government. A renewed attempt to install a regional form of government failed to attract sufficient popular support in 1998, leaving a strong dual structure of territorial governance. Concerns with national identity outweighed regionalism in the national referendum, reflecting the absence of an established tradition in regional governance. Nevertheless, the central government sought to fall in line with established EU practice in regionalization and thus adopted a 'dirigiste' approach and implemented a new regional tier of territorial government in May 2000. The desire for a supra-local, subnational form of territorial government was particularly pronounced in the capital region, Lisbon, seeking to 'consolidate and improve its position as a major European and world city' (Syrett and Silva, 2001: 9). Having to rely on collaboration and cooperation between 19 local authorities covering the Greater Lisbon area, suggests a situation not dissimilar to that of Greater London until the installation of the mayor and London Assembly. Interlocal cooperation in the Greater Lisbon area was statutorily formalized in 1991 as Metropolitan Government. This term, however, suggests more power and responsibility than the institution actually holds. Being little more than an association of local authorities, with no direct democratic mandate, its democratic legitimacy comes from local authority councillors delegated to the metropolitan council whose main role is strategic metropolitan planning and control, the provision of infrastructure and environmental issues. This arrangement is not dissimilar to that found in other European countries. Councillors lack a direct political mandate, and local interests are inevitably challenging the willingness to engage in regional compromises. Political stalemate leads to accusations of the council being little more than a talking shop (Syrett and Silva, 1999). It is against this background that central government wants to push regional governance towards a more responsive, business friendly and collaborative approach by using regional development agencies as the main vehicles for delivering regional governance, largely independent of local political conflicts.

In the Lisbon area, there are two private sector-style development agencies each covering about half of the Lisbon region (Lisbon Metropolitan Area): one for the southern half, including nine local authorities of Greater Lisbon, and five of the adjacent Setúbal district (NUTS 3). The origin of this bipartite division is essentially political, based on strong ties with the Communist Party in the Setúbal region in the 1980s (Syrett and Silva, 1999), which led to a separate grouping of local authorities. Regional boundedness is thus not born out of regional interests and the realization of specific advantages from collaboration, but rather from political comradeship. The second development agency, covering the northern part of the Lisbon Metropolitan Area, is essentially a marketing tool for the municipality of Greater Lisbon (NUTS 3) with emphasis on public–private partnership. Created in the early 1990s, it is a construct based on a single urban authority and does not therefore reflect regional economic linkages and interests going beyond the Greater Lisbon authority. Both agencies thus essentially represent bottom-up regionalization, but with little evidence of a genuine input of, and response to, regional issues of economic development and linkages, and resulting policy-making requirements. They do, however, reflect a clear shift from (centrally-defined) government to a broader form of governance, including the private sector and other interest groups. It may be through them, and their specific links beyond the (local) boundaries of the two agencies, especially the one of the Greater Lisbon council, that a regional territorialization may emerge that responds more genuinely to economic and political links in the Lisbon region. It seems that an established separation between historic reliance on top-down central government regulation on the 'one side', and informal, non governmental local problem management, need to be overcome and merged into a form of genuine cross-sectoral governance of city-regional affairs in development planning.

The Randstad Holland: state-inspired polycentric city region

The Netherlands have become well known for urban-regions through their project of the Randstad conceived as a concept in the 1930s in conjunction with the development of the new Schiphol airport (Lambooy, 1998). The two city regions of Amsterdam and Rotterdam dominate the northern and southern sections of the Randstad respectively (Map 4.2). Consisting of a ring of (individual) cities around a difficult to build on green area (including the airport), the 'Green Heart', it was very much a planning concept aimed at development control and facilitating a more even spread of urban development rather than allowing individual cities, like Amsterdam, to take the lead (Dieleman and Faludi, 1998; de Regt and van der Burg, 2000). Netherlands' spatial planning has emphasized a polycentric structure, with the Randstad being the most important agglomeration and example of this approach. The Randstad's polycentric nature was seen in the 1960s as a unique advantage a major metropolis (SPESP, 2001). Given its background, the Randstad region is not a functionally integrated urban structure (Lambooy, 1998), but is kept together by the common economic attraction of the airport and Rotterdam 'seaport'.

The Randstad, like the Rhine-Ruhr region and other polycentric metropolitan regions, is seen as an important vehicle for boosting the participating cities' chances of entering the level of 'global cities' in the international competition for business investment and, especially, headquarters. 'On their own, cities like Amsterdam . . . can never hope to compete with true world-cities like London, New York and Tokyo for high-level functions and headquarters of large corporations' (Dieleman and Faludi, 1998: 373). This aim is reinforced by the current spatial development goals for the region, seeking to further the Randstad region's function as a centre in the north-west European economic networks (de Regt and van der Burg, 2000: 694). Expanding the Randstad into the wider Delta Metropolis project serves this objective, with its name highlighting the metropolitan rather than city ('stad') oriented focus. Nevertheless, the two main cities, Amsterdam and Rotterdam, 'always profile themselves as separate entities, not as a part of the Randstad' (Dieleman and Faludi, 1998: 370). This reflects the inherent localism and competitiveness among the cities in this city region, seeing themselves as *primi inter pares* and seeking to keep any competitive success entirely to themselves. Localist competitiveness also shapes the relationships between the core cities and the smaller suburban communities, because of the financial implications of population shifts (e.g. tax revenue). These tensions are, so far, not deemed to be adequately addressed by the system of national and provincial governments, and this has led to new discussions

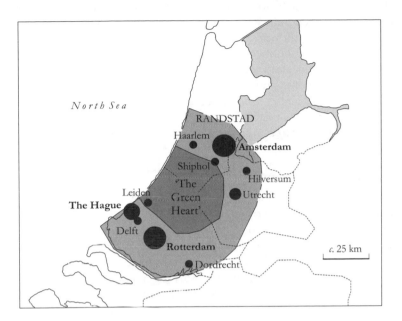

North Sea

RANDSTAD

Haarlem

Amsterdam

Shiphol

Hilversum

Leiden

'The
Green
Heart'

Utrecht

The Hague

Delft

Rotterdam

c. 25 km

Dordrecht

Map 4.2 Randstad Holland.

about inserting another 'intermediate' regional scale of governance to address financial and governmental imbalances (SPESP, 2001).

From a national perspective, the Randstad is of great importance to staking a claim in the globalizing economy. As such, it is an integral part of spatial planning and policy. This includes a distinct shift in paradigm from growth management (mainly housing) to a pro-active support of economic development, including international competition. The Fourth Report on Spatial Planning made this effort in 1988 against the background of a sluggish national economy, and stressed the competitive advantages of large urban areas (Priemus, 1998). This attempt at shifting the agenda needs to be seen against the background of a Germanic-style complexity, comprehensiveness and effectiveness of Dutch planning (Hajer and Zonneveld, 2000). More recently, changing social and economic circumstances, especially the shift towards international competition of urban regions, challenges the strong emphasis on institutional comprehensiveness, with its limited love for informal and thus less visible arrangements.

The complexity of extensive inter-governmental coordination, based on a general culture of consensus when dealing with divergent interests (Hajer and Zonneveld, 2000) is similar to that in Germany. The main difference involves the absence of a strong regional level of government. As a result, 572 local authorities are directly juxtaposed against the national government.

There are 12 provinces which nominally mediate between the two tiers of government, but they lack administrative and governmental power, especially vis-à-vis the main urban authorities. Referring to these inherent weaknesses, the provinces are also being dubbed as 'invisible government' (van der Veer, 1998: 27). Strong urban localism, particularly in the main cities, and a reluctance by the state to devolve power, including financial revenue, has contributed to the absence of an effective tier of regional government. This makes the management of urban regions such as the Randstad more difficult in the attempt to respond to an increasingly more competitive regionalism. Thus, there is municipal fragmentation within the metropolitan areas, with local authorities fiercely protecting their autonomy and local identity. This situation made it politically unfeasible, as part of local government reform, to aggregate local government permanently into larger metropolitan entities. At the same time, the national government rejected the idea of establishing a further, fourth tier of administration, claiming added bureaucracy, but essentially fearing another political power base potentially challenging the centre's competencies. Nevertheless, an institutionalized response to economic regionalization had to be found, because informal arrangements on the basis of single-purpose regional cooperation between local authorities seemed 'messy' and undemocratic. One proposal in the mid-1990s was to create larger city regions (or city provinces) as 'extended' local authorities around the main urban centres, thus effectively recommending a series of monocentric city regions as the backbone of development planning and policy. Political resistance, especially from the threatened suburban authorities, however, led to an abandonment of this proposal, highlighting the essentially political rather than functionally driven nature of the debate which was shaped by 'the changing political stances of the actors involved and by rivalry between and within government layers' (van der Veer, 1998: 32).

With a growing shift from technocratic government to a more inclusive (democratic) form of governance and thus a growing number of actors, interlocal rivalry is likely to be increased. The role of individual personalities in managing such coordination can be illustrated by the case of Amsterdam's efforts to expand into the region. Initially, the suburbs were unwilling to cooperate, seeking to maintain their identity and autonomy. It was only with the newly elected mayor of Amsterdam towards the end of the 1980s that new political bridges were built and communication networks established between Amsterdam and the neighbouring localities. The mayor also accepted a numerical minority stake in the new regional council that comprised delegates from the 15 participating local authorities, and thus underplayed, at least formally, the dominant role of the central city. The statutory basis of the collaboration was the Joint Agreements Act which sets out informal collaborative arrangements between munici-palities within a formal framework. The original aim of the cooperation was to establish a formal and territorially defined city province as 'official' city region with an integral regional government, instead of the established much looser arrangement for the region with groups of separate local authorities. In effect, however, discussions about such a political-governmental regional entity led to second thoughts among the participating suburban local authorities and to the departure of some for fear of a looming loss of power by stealth. They were interested in a regional umbrella organization with few powers to offer a platform for informal consultation and coordination of local policies, but were wary of any more far-reaching ambitions. The mere attempt at institutionalizing a formal city-regional government was felt as a threat rather than an opportunity, from which, when completed, there would be no return. Informal, less institutionally binding arrangements of regional governance were, by contrast, accepted, as they seemed to offer the best of both worlds: local policy-making autonomy and identity, and the added political muscle of a regional voice. The pressure to place such arrangements into a fixed framework of established formal arrangements for planning and government appears to have been counterproductive. Such reservations seem to matter less where there are fundamental economic problems which seem to demand a joint response as the only viable option. The sense of a common fate and common historical and cultural background seem crucial for the willingness to collaborate between the central city and the surrounding suburban authorities.

Effective regionalization, carried by the participating local authorities, seems also to be hampered by the continued strong role of the central state in subnational governance through its financial control. Effectively,

> the amount of centralisation in the Dutch tax and grant system makes the discussion of metropolitan government in the Netherlands entirely different from the discussion in many other countries. In fact, if one compares the scale of the Netherlands with other countries[,] one might as well conclude that the real metropolitan government is the national state itself
>
> (van der Veer, 1998: 47)

This allows the state to continue its traditional redistributive role and co-ordination between government tiers rather than facilitating a genuine integrated urban-regional government for the metropolitan areas. The insistence on formalized, institutionalized structures has encouraged second thoughts among participating actors. Encouraging and accepting less finite and more flexible informal arrangements, despite their perceived lesser democratic nature, may have been more successful in establishing required urban-regional governance. This became clear also during the attempts to establish formalized and institutionalized transport planning regions in addition to existing administrative territories in the early 1990s. Powerful, directly elected transport authorities were to be set up in metropolitan areas in collaboration with the relevant urban authorities. In all other areas, inter-communal cooperation was meant to deal with region-wide transport, after responsibilities had voluntarily been transferred from the local authorities to the new regional transport body. However, there was no clear procedural mechanism for these changes, and local authorities proved reluctant to transfer power. Consequently, the initiative lasted little more than a couple of years.

Against this background, a new approach of establishing single-purpose and temporary collaborative regional arrangements, so-called ROM regions (Spatial and Environmental Planning) (Kerstens, 1998), has been adopted. Any transfer of local responsibilities will be reversed after the set task has been completed by the regional body. This ensures the local authorities of their independence and removes the anxiety about a possible permanent loss of local powers. Municipalities are thus more willing to cooperate in the interest of regional concerns. The experiences in the Netherlands demonstrate the challenge regions from the established powers of central state and local government. Both are reluctant to engage in any concessions towards regional governance, which may lead to a permanent loss of powers. In the absence of a clear and formally established representation of regions in government, voluntary collaborations across and between non-regional government tiers appear to be the only option available, because they offer a visible escape route for collaborators to allay their concerns about a possible permanent loss of powers. It seems that, in a centralized state, unclear responsibilities in regional matters heighten a sense of risk about potential losses in policy-making capability as a result of engaging in collaboration. This problem includes both mono- and polycentric regions where municipalities are vying with each other about potential supremacy in the first case, and are wary of the dominant city's position, in the latter.

Regions and regionalization in unitary states: caught between central and local competition for power – summary

The experiences with regionalization in the unitary states have shown a general shift towards recognizing the importance of regions as a separate scale of government in their own right. For the strong central states this resulted in an inherently ambivalent attitude towards regions, because there is little appetite for losing power to an additional, lower tier government. At the same time they recognize the need to address the challenges of inter-regional competition, also encouraged by the EU's spatial policies. In all examples, attention clearly focused on cities as centres of economic development and thus as regional growth centres, particularly in monocentric regions. There is little appetite for new territorial-based regionalization, but for a network-based and less formal approach with a lesser challenge to existing governmental structures and power distribution. This suits the local level of government as well, because they, too, are less than enthusiastic about additional players in the government sphere who pose a potential challenge to their status and policy-making capability. A loose, non-committal, informal arrangement is their preferred option, particularly when competing with each other in the same region. In polycentric regions, in a unitary state, new territorial regionalization with a formal tier of government faces opposition and hostility from both local and central government. 'Soft' informal, voluntary options thus seem the only realistic option for addressing regional scale issues. Whether a more federal state structure with inherently stronger and secured positions for subnational government makes any difference to the nature of collaborative arrangements will be investigated in the following sections.

Regions in federal and quasi-federal states

The examples in this section explore mechanisms of regionalization in a more devolved, federal state structure as external framework and look, as above, at polycentric and monocentric regions. They show clearly that a unitary, centralized state system triggers some insecurity among subnational governments about their position and policy-making capacities, which, in turn, affects their willingness to collaborate in regional matters. Such collaboration at the local level may, for instance, include the transfer of some powers to a regional body as a joint local initiative. The fear of losing such powers for good, in the absence of a statutorily protected status and thus the possibility of being able to repatriate any surrendered powers, hampers local engagement in collaborative regionalization. It will thus be interesting to see whether this situation will be different in a federal state structure with protected secure local government rights.

Regional and new metropolitan renaissance in Italy?

Italy has gained particular attention in the debate about regions and regional policy since the 1980s through the case of the 'Third Italy' (see Chapter 2). The institutional support of informal business practices has been argued to have been crucial in helping to turn the Third Italy into a new paradigm for successful regional policy. The 'Third Italy' became a cause célèbre for successful regional economic development and the importance of 'appropriate' regulation. Given the structural, cultural and historic divisions between the Italian north and south, regional policy operates quite differently in the less industrialized, economically weaker *mezzogiorno*. The economic dependency of the south on the northern economy has created a matching dependency of regional governance structures on the national political system and the associated (changing) networks and power relationships (Martinelli, 1998). These divisions have stimulated the current political and, especially in the north, popular shift towards greater devolution to the 20 regions. In response, these centrally defined former planning regions have, for the last 10 to 15 years, gained political clout and policy-making capacity from the centre. Beside the regions, the provinces and municipalities are the dominant actors in subnational government. The most fervent supporters of greater devolution have been the northern regions whose leadership came from the same political background, which facilitated their close collaboration in lobbying national government. The southern regions, meanwhile, have been less enthusiastic, fearing initially that more regional power meant more immediate direct control of local matters by the state (interview with officer of Emilia-Romagna regional administration in Bologna, 19 June 2001).

Economic changes during the 1980s contributed to the reassessment of the traditional Keynesian state-managed redistributive policies (1950–80) to 'modernize the mezzogiorno'. While for the industrial north this meant readjustment and restructuring with new practices in governance (Third Italy), this option was not so readily available in the south, because of its economic dependencies and insufficient indigenous economic potentials. EU structural policy offered some support to these dependent areas, together with attempts to 'regionalize' regional policy through devolving responsibility, primarily for regional planning, to the regional governments. This was in response to the general paradigmatic shift from 'big' regional policy through state intervention to small-scale and more targeted 'indirect micro-intervention' (Martinelli, 1998: 18).The 'Third Italy' model seemed to fit the bill, and its features served as examples of modernizing regional policy elsewhere in the EU.

These changes were relatively short-lived, as pressures from EU integration grew. The collapse of the Eastern Bloc, and severe financial cutbacks in the wake of joining the Euro, brought an end to established forms of policy-making in favour of neo-liberal policies. This also meant the end for the state quango for regional development, Cassa del mezzogiorno (Gualini, 2000). Instead, since the early 1990s, EU Structural Fund-inspired criteria and principles of territoriality have been used, with a shift towards regional *governance* by involving multilevel government and non-governmental actors. But there has been no coherent overall regional development plan (Martinelli, 1998). Network-based relationships are now being encouraged, following the paradigm of the Third Italy model. This includes acknowledging the role of local, especially urban, forces in regional development, as illustrated by the case of Turin.

The city of Turin has developed a comprehensive strategy, 'Torino Internazionale', to promote the city as an international locality. This includes promoting the city as 'a European metropolis', 'resourceful', and acting as 'decision-maker' (City of Turin, 2000). One of the six strategic 'paths' (Table 4.2) projected is to 'construct a (new) metropolitan government', including a Metropolitan Conference as policy-defining negotiating platform (not unlike the German model of Regional Conferences), to govern the whole metropolitan area. The regionalization process seeks to address the functional entity of the Turin urban region and adjust the territorial government accordingly. But there are no formal boundaries for the metropolitan

Table 4.2 Turin's internationalization strategy, 'Torino Internazionale'

6 Strategic avenues	20 Strategic objectives	84 Policy initiatives
1 Integrate the metropolitan area into the international system	e.g. • foster international cooperation • improve accessibility to city	e.g. • create a standing committee for international cooperation • improve airport and its accessibility
2 Build a metropolitan government	e.g. • create a new form of urban governance • develop services for the whole of the metropolitan area	e.g. • establish a Metropolitan Conference to govern the complete metropolitan area • create a transport agency to manage traffic
3 Formulate a strategic development programme	e.g. • build a polytechnic university of international standing • facilitate professional development and qualification of workforce	e.g. • develop areas of scientific excellence in the polytechnic • develop a pilot project of collaboration between academia and business
4 Foster economic development and employment	e.g. • create an entrepreneurial 'climate' • develop employment related policies	e.g. • develop a 'technology district' • foster technology transfer to business
5 Promote Turin as city of culture, tourism, commerce and sport	e.g. • improve cultural programme • place Turin on the map of international business tourism	e.g. • improve city's museums • develop a local/regional product image 'made in Turin/Piemonte'
6 Improve urban qualities	e.g. • regenerate city centre • promote Agenda 21 as part of city's development strategy	e.g. • regenerate run-down parts of the city • improve environmental quality (pollution)

Source: Based on Turin City Council (*c.* 2001): Torino Internazionale – Strategic Development Plan' (unpubl.).

region. A functional delimitation of the proposed region may well cut across existing administrative 'official' boundaries. In the case of Milan, for instance, there are five formal provinces with 187 municipalities within its functionally derived boundaries (SPESP, 2000) which would need to be regrouped by such an additional metropolitan 'designer region'. Currently, general regional framework plans are the only formal instrument available to address developments at the metropolitan regional scale.

This city-focused regionalism has been highlighted by Lefèvre as a general 'renaissance of metropolitan governments' in the new regionalism of the 1990s (Lefèvre, 1998). In 1990, Italy's legislators formally attached a regional dimension to the city by establishing the *città metropolitane* as a theoretical construct analogue to English metropolitan authorities. Although never really properly implemented, this top-down creation was somewhat overtaken by state sponsored, locality-focused regionalization initiatives. These include the 1994 creation of a formal urban network within the Province of Bologna as an expression of the new locality-based nature of regions: the 48 municipalities of that province signed a contract of cooperation with the region for a range of specified regional matters which would be dealt with as required. This arrangement was therefore not an *ex ante* redistribution and formalization of powers, but a problem-specific and time-limited negotiated, informal agreement for a clearly specified territory as sum of the participating municipalities' areas. The objective of the arrangement is to constitute the metropolitan city as a operational entity, defined by its functions and (temporary) tasks, rather than a new form of top-down implemented institutional scale of government. In the Bologna model, regionalization is to be flexible, task-specific, interlocally agreed and voluntary, rather than state defined. The Metropolitan Conference represents the visible political-institutional expression of this arrangement, comprising the relevant localities' mayors and the President of the Province (as in Germany's Regional Conferences). It is essentially a platform for discussion and negotiation between local representatives, with local councils remaining the actual democratic forms of government also at the regional level. Policies concentrate on three areas: territorial (transport, environment, planning), administrative-financial (government) and social services (service provision). The arrangement of the Conference allows problems and policies to be focused on small projects, which are easier to solve in a pragmatic rather than general political way.

In effect, therefore, regions in Italy may be seen as little more than aggregations of individual localities which act as the centres of *actual* decision making and provide the grounding in civil society. Regions are effectively the result of regional clustering of a 'multitude of local societies which have their networks, strategies and cohesion at the municipal and the provincial levels' rather than a genuine, separate regional scale of government (Bagnasco and Oberti, 1998: 162). As a result, there is no genuinely regional interest or dynamism to translate into region-based policies and networking within the state system. The subsequent absence of corresponding regional institutions and forms of governance hinders the development of a *region*-based civil society and thus regional 'civicness'. This makes the development of appropriately scaled, responsive, credible and thus 'efficient' territorial regions more difficult (after Putnam, 1993, referred to in Bagnasco and Oberti, 1998: 150). This situation may, however, change with the now immanent constitutional changes to boost regional capacities. The critical point is here, as elsewhere, the financial situation, and thus the degree of *genuine* power devolution to the regional scale. Regardless of the changes, the state does retain some directive role to maintain coherence in the overall administration. Thus, there are regular state-regional conferences convened by the state government in Rome, and there are also local–regional conferences hosted by the regions to discuss collaborative work between the regional administration and the relatively powerful municipalities. It is in these conferences that, usually politically-based, alliances and collaborations emerge. Similar collaborative alliances exist between the regions against national government. It thus remains to be seen to what extent regions can develop their own identity and policy agenda between central state and competitive and unenthusiastic local authorities.

Culture and history driven regional devolution of Belgium

Regionalization in Belgium is a somewhat special case, because of its internal cultural-political divisions, institutionalized by the two semi-autonomous regions of Flanders and Wallonia, and the Brussels capital city region. In effect, these divisions have resulted in Belgium changing from a unitary to a federal state over the last 25 years. The Belgian constitution distinguishes two types of federal entities: 'communities' as cultural-linguistic groupings with a less clearly defined territoriality, and 'regions' as the clearly defined territorial divisions at the regional scale. It is the three regions that carry responsibility for most domestic state matters, with international functions left to the federation (SPESP, 2001). The deep divisions between the three regions are reflected in the fact that there are no platforms installed for interregional dialogue and collaboration. Within the regions, there is no distinction between urban and non-urban municipalities, thus somewhat ignoring the different roles of the two in regional economic development. The rationale for that approach may have been an attempt to create somewhat of a level playing field for intermunicipal associations within the two main cultural regions as voluntary arrangements for supra-local service provisions. These associations are 'open access', i.e. local authorities can join and leave as they wish, to serve their best interest. Such informal collaboration has also been possible for the three cultural regions since 1988, driven largely by the realization that jointly they may achieve greater international competitiveness. There is scope for functional complementarity between the regions' uneven degree in urbanization. There is no major urban centre to compete at the global city level outside Brussels. Instead, an 'unambiguous urban pattern' exists with 'diffuse developments in diverse directions' (Albrechts, 1998: 413). To overcome this disjointed urban structure, and tie in with the international standing of Brussels, a strategic framework of the 'Belgian Central Urban Network' has been developed to include, and link up, all the main cities in the wider city region of Brussels, encompassing all three regions.

Devolution and a growing awareness of international competition for corporate investment encouraged a review of the traditionally strong focus on physical planning and control within a rigid, firmly institutionalized system. This appeared increasingly insufficient. Greater strategic policy input and moves towards area marketing were recognized as being of crucial importance. In the absence of an obvious regional direction of development for Flanders, a collaboration of the main cities as an 'urban region' was envisaged as a way forward (Albrechts, 1998). They include Antwerp, Ghent and Brussels. The Randstad Holland was seen as the main example to follow and thus establish an internationally competitive polycentric region, although with Brussels clearly as *primus inter pares*. The project was pushed by the central government concerned about high unemployment and hoping to achieve economic growth through greater visibility as a globally relevant metropolis. The concept is thus clearly imposed top-down, with little evidence of indigenously developed affinities between those cities. Instead, the model has been established as a policy response to global competition, not because there was a genuine urban network as the basis of a region. In effect, a multicentric region (Flanders) and the monocentric Brussels region have been partially superimposed to extend the capital region's international standing into the comparatively lower positioned Flanders city network. The new concept serves also as a means of international representation of Flemish identity, which is not recognized in an EU institutional context (Dieleman and Faludi, 1998; Zonneveld and Faludi, 1997) and as such has a greater than merely planning-functional role. To what extent such an artificial construct established from outside the designated region can work has to be seen.

Multitier regional government, polycentric regions and devolution in Spain

Spain has historically been a centralized state, but over the last two decades has seen a growing shift towards a more federal arrangement, because democratization and decentralization were seen as going hand in hand (Rodriguez-Pose, 1996). Devolutionary pressures were particularly high in Catalonia and the Basque region. The situation has some similarities with the one in Belgium as discussed above. In Spain, however, different paths towards devolution are constitutionally possible, with historic regions allowed a faster process than other regions wanting to gain greater autonomy

(Rodriguez-Pose, 1996). There are national, regional metropolitan, regional and municipal spheres of government, all of which interlinked at the regional scale. The regionally oriented approach is not made easier by this rather complex administrative structure with an institutionally 'thick' meso-level between local and national government: there are the main Spanish regions, such as Catalonia, then four provinces and 41 comarcas (counties) as groupings of municipalities. Planning and development are the responsibility of the top level regional bodies (e.g. Catalonia) and the local level, thus potentially confronting quite different perceptions of regional scales, perspectives and interests.

The tiered system of spatial governance is illustrated by the Madrid region, or Community of Madrid: there are the municipality as the core city, the metropolitan area, and the rest of the Madrid region. All are rolled into one single-province community, i.e. city region. Its responsibilities encompass those of its constituent parts, i.e. government tasks at all scales within the capital region's boundaries, but the tasks remain with the respective tiers of government. Responsibility for regional planning rests solely with the Community as representative of the Madrid region. More contentious is the shared responsibility for local town planning, because this requires inter-scalar collaboration between the municipalities and the Community of Madrid as strategic body. This collaboration is formalized through two coordinating bodies, the Commission for the Coordination of Regional Action and the Council for Regional Policy (SPESP, 2001). Both offer platforms of consultation and negotiation between the various local actors in the region. In addition, there are non-governmental consultative bodies, e.g. the Madrid Federation of Municipalities, all of which contribute to the formulation and modification of the regional development strategic plan.

This institutional complexity of spatial governance is also reflected in the Barcelona region and the growing scale of governance in response to the centrifugal functional and population shifts: Barcelona city council, surrounded by the immediate hinterland of Barcelona County, and then the wider Barcelona city region (Burns and Cladera, 2000). This arrangement is comparable to that of the Madrid region. Despite this administrative complexity, which highlights the unclear scalar concept of 'region', no indication is given of how this new envisaged metropolitan region is to be delimited. As Burns and Cladera (2000) point out, taking merely the administrative boundaries of the currently responsible Strategic Planning Authority would include 308 municipalities with an area more than twice the size of the relevant current urban region. Alternatively, a functionally-based delimitation could be adopted that would be smaller in diameter, but would cut through established administrative areas. Such an approach is currently being discussed out as a possible model for the governance of south-west European urban regions. On that basis, the new Barcelona Metropolitan Statistical Area incorporates 217 municipalities, with some 4.4 million inhabitants.

Having lost its metropolitan government in 1988 after a struggle between city council and regional government (Morata, 1997), the challenges and perceived opportunities of staging the Olympic Games in 1992 encouraged wider political networking by Barcelona's city government and especially its mayor, which was extended well beyond the immediate geographical area of the Games. Since its success in hosting the games (Marshall, 1996), Barcelona has repeatedly been hailed as an example of good practice of metropolitan governance and post-industrial transformation (Burns and Cladera, 2000). Since then, a succession of major strategic plans – Barcelona 2000 (Marshall, 1996) – has highlighted the importance of networks between key organizations and personalities, which go beyond city limits and extend into the wider city region. The first Barcelona 2000 Strategic Plan (1988–1993) explicitly states as one of its aims the linking up of Barcelona with European and world city networks, which affects the ambition to break out of its provincial (peripheral) embeddedness and gain global recognition. This ambition was re-emphasized by the second incarnation of the Plan (1994–1999), stressing as its overall aim 'to increase the integration of the Barcelona region into the international economy' (Strategic Plan, quoted in Marshall, 1996: 156). This ambition affects Barcelona's new understanding of itself as embodiment and representation of the wider city region and its development, and not merely of the immediate administrative urban area.

The current third Barcelona Economic and Social Strategic Plan for 1999–2005 identifies four strategic development models with varying local (metropolitan) and regional considerations. One of the models envisages a polycentric urban region consisting of a group of functionally interrelated cities which would accommodate almost three quarters of Catalonia's population of some 6.5 million. Barcelona would, inevitably, gain a special, primate status within the regional system of cities with their varying centrality. The Plan expressively refers to a 'region of cities' whose collaboration in the regions would add extra capacity to the area's (and Barcelona's) competition in the global economy, because the whole region would offer more than the sum of its individual local components (Burns and Cladera, 2000). In adopting a more explicit regional perspective, the strategy responds to clear indications of a continued decentralization of population and economic activity from Barcelona into the region, thus effectively expanding the city into the region and with it the scalar perspective of related challenges to policy making. This would reinforce the polycentric nature of the region with its reduced hierarchical differences among the leading cities, while enhancing Barcelona's competitive standing among the other major city regions in Europe and, so it is hoped, beyond.

Despite ongoing regionalization and some devolution towards the five main regions, the centre maintains considerable influence over planning and development through infrastructure investment decisions. One of the main characteristics of the Barcelona model is its openness and collaborative nature

(Marshall, 2000), thus encouraging governance rather than government. This openness and inherent flexibility is based on a functional-political rather than territorial definition of 'region'. It was thus a policy issue rather than an administration-driven territorial form of regionalization. Such a shift towards 'designer regions' (see Chapter 3) is now increasingly seen as a more flexible and issue-responsive form of regionalization, not only at EU level (with support from Structural Funds), but also in other countries like Germany. In the case of Barcelona, this approach was favoured, as it did not require the (often) cumbersome installation of new institutions, but could operate through the existing structures in favour of the metropolitan region as a whole. There was no general shift towards formalizing such an approach, but there was a clear reference to Barcelona as a 'metropolitan region' ... 'with an organizational capacity which will facilitate economic opportunities to its citizens' (Marshall, 2000: 11). The third version of the Barcelona regional plan adopts a dual role of project management and implementation, and strategic planning and policy making. The absence of a single regional body, and reliance on cooperation between and within government tiers has kept the structures for regional policy making flexible and open to new actors so that changing policy requirements can be met. The inevitable negotiation processes, however, may slow policy-making processes down and thus be less effective.

Barcelona has not only been a major driving force in the regionalization in north-east Spain, but also has engaged actively in the European networks of city regions. As the 'most important metropolitan area of the Western Mediterranean Arc' (see Morata, 1997), the leaders of the province of Catalonia and Barcelona were instrumental in driving the establishment of the Committee of the Regions at the EU. The European reach of Catalan initiatives include the pursuit of transnational regional collaboration, such as between the 'Four Motors of Europe' regional grouping, including the four main 'innovation regions' Baden Württemberg, Catalonia, Lombardy and Rhone-Alpes (Morata, 1997). Most of the cooperation is informal, involving a range of non-governmental actors, such as chambers of commerce. The city of Barcelona has also been a main driving force in establishing the 'Eurocities' network consisting of 97 cities from 26 countries, with 20 cities possessing associated status as non-EU cities. The main focus of the network is to highlight and politically underpin the major role of cities as centres of innovation and development (Morata, 1997). The third main regional network which comprises Barcelona is geographically more integrated and coherent, concentrating on the western Mediterranean, including Montpellier, Palma de Mallorca, Zaragoza, Toulouse and Valencia. Much of this collaboration was economy driven, to address the wider regional impacts of the 1992 Olympic Games. This greater economic interrelatedness, and the cultural ties, make this regional network more immediate than the EU-wide C 6-network.

The example of Spain, and Barcelona in particular, clearly illustrates the multilayered nature of formal territorial government in city regions, and the role of informal networks, each with their particular range of members, aims and objectives. Collaborative arrangements, either territorial or functional by nature, are of varied intensity and geographic reach, and may be altered in their composition simply by a change in membership. The case also shows the importance of cultural-historic identities in pressing for federal regionalism in a centralized state, similar to the Belgian case. These processes overlay the differently scaled, more localized regionalization in individual city regions and the much broader EU-wide regionalization through city networks, illustrating the multiscalar nature of regionalization in Europe. The seeming dominance of institutional territorialization should not, however, as shown in previous cases, obscure the crucial role of individual personalities and strong overall policy objectives as beacons of common regional interest and crystallization points of region forming.

Centrally sponsored project-driven regionalization and localist competition in a polycentric city region: the Ruhr region in Germany

The federal structure and regional planning and policy-making arrangements in Germany have been discussed in detail elsewhere (Chapters 5–7). Here, the focus is on two brief examples of mono- and polycentric regionalization in two German *Länder* with long-established federal practice as part of our comparative synopsis across the EU. They are the polycentric Ruhr region and the monocentric Frankfurt region, respectively.

The first example looks at regionalization in the polycentric old industrial Ruhr region in northwest Germany as part of a regional rebranding exercise. Established in 1989 for a 10 year period, the *Land* inspired Emscher Park International Building Exhibition (Internationale Bauausstellung Emscher Park) was part of the *Land* (North-Rhine Westphalia) government's new emphasis on region-based interlocal cooperation. The 'region' was a loose concept of functionally delimited territoriality rather than a fixed part of the hierarchy of territorial government. Placing this new approach to region building outside the established governmental hierarchy was also a deliberate attempt at bypassing the bureaucratic, multilayered structures of planning as the traditional way of (enforced) intercommunal cooperation (Danielzyk and Wood, 2000). The Ruhr region is a good example of considerable 'institutional thickness', comprising municipalities, government office regions, several inter-municipal associations for specific tasks, several economic development agencies and non-government organizations. This complex web of regional actors had no common regional development vision.

The exhibition was intended to demonstrate 'good practice' in redeveloping and rebranding old industrial regions. The Emscher Park concept aimed at bringing together 17 urban authorities along the Emscher River as an industrial corridor across the northern part of the Ruhr region. The nature of IBA as an exhibition of themed individual projects, based on a common strategy, reflects the dual approach to regionalization: formal, locally-based planning regulation, and new strategic policy-making. Developing the strategy, management and master plan was the responsibility of the new private sector style exhibition planning company, while the actual implementation of the individual physical projects was, in line with statutory provisions, the responsibility of the local authorities involved. In effect, localist tendencies had to be consistently counteracted, because project implementation depended on local support. The physical evidence of successful cooperation in the shape of local building projects was helpful, as it allowed the relevant local policy makers to claim achievements and results for their localities.

The project exemplifies an open, flexible, voluntary and issue-based collaboration as the basis of regionalization and as such marks a deviation from the traditional formal planning hierarchy as the basis of regions. The single, overarching aim of the IBA project has been to turn a derelict industrial area into an environmentally attractive, greened urban living space with an underlying theme of industrial heritage. Such open entry, and exit, regionalization appears to be the only option in a region consisting of some 20 neighbouring urban municipalities which are fundamentally in competition for new business investment. Inevitably, localism is never far away when it comes to balancing local and regional economic policy objectives. This localism is further supported by historically strong local identities of steel-making and mining communities (Danielzyk and Wood, 2000). There has, therefore, been decreasing willingness to cooperate due to growing economic problems in the region and thus growing interlocal competitiveness for whatever investment potential existed. This local competitiveness undermined the long-established institutionalized form of (voluntary) inter-municipal cooperation in the shape of the Regionalverband Ruhrgebiet (Regional Association Ruhr), bringing together the more than 50 local authorities across the Ruhr region in a municipal association with no independent statutory powers. Its main aim has been the protection and management of open space in the region, and, lately, image marketing and research. Its disbandment is imminent, with the intention of the *Land* government to replace it with a more centrally managed regional development agency.

After ten years of operation, some deficiencies of the incentivized, informal regionalization approach have become apparent. They include the credibility and acceptance of regions as real, rather than merely opportunistic constructs in response to financial incentives. The Ruhr region benefited from an existing image as a territorial entity. Another weakness is the less inclusive form of governance, favouring a continued use of elitist representation through established 'official' channels of (local/urban) government, especially key individuals and their connections within the political system, in particular the ruling left-of-centre party, the SPD. A vigorous defence of local autonomy makes formal and thus more practical changes to established practices and relationships between actors, difficult for fear of a dilution of local policy-making autonomy. Nevertheless, new forms and practices of regional policy making have been explored, encouraged by financial incentives from the *Land* government, offering alternative routes to regional governance outside the formalized, technocratic hierarchy of territorial government. Such alternatives have also been attempted elsewhere in Germany, such as in the monocentric Stuttgart region, to achieve greater flexibility and problem relevance of policy measures. There, under the leadership of Stuttgart as the dominant city, the Stuttgart Regional Association was formed in 1994 to overcome intercommunal competition and provide a joint platform to negotiate common agendas and policy measures. Economic development, environmental issues and transport infrastructure are the main concerns (SPESP, 2001). Similar to the Ruhr region, other informal, regional alliances and interest groups have been set up, including the Stuttgart Economic Development Corporation as a public–private organization comprising over 100 municipalities of the region, and various regional marketing agencies.

Institutionalized versus informal regionalization in the monocentric Frankfurt region

The regionalization or, more accurately, the re-regionalization process in the Frankfurt (Main) area illustrates particularly clearly the inherent tensions between necessary cooperation and continued competition between local and regional interests. The Frankfurt region was one of the first to engage in locally based regionalization some 25 years ago in response to the rapid suburbanization processes in a successful city region. The Frankfurt City-Region Association (Umlandverband Frankfurt) brought together the city of Frankfurt and the surrounding suburbanizing local authorities in a formal arrangement with fixed territorial boundaries, as defined by the areas of member municipalities (Table 4.3). These transferred some of their responsibilities to this association whose main role was, as in many other conurbations of this kind, public transport and waste management as the two services that link city and region most visibly. The Association, however, had few real powers and was essentially too small in size, covering only parts of the functionally interconnected Frankfurt region (Bördlein, 2000). Local authorities were reluctant to transfer any of their statutory powers to the new organization and remained half-heartedly committed to this arrangement. The more peripheral local authorities in the region, in particular, also viewed the Association with suspicion as a likely tool of encroaching domination by the city of Frankfurt. Furthermore, the region lacks a common identity, because there is no common historic root, but merely current

Table 4.3 Models of regionalization in the Frankfurt (Main) area

Characteristics	City-Region Association Frankfurt (Umlandverband) (abolished) 'metropolitan model' (one administration for all tasks)	Formal region at county level (Regionalkreis) (proposed)	New regionalization with 3 components: Planning Association, Council of the Region, 'intermunicipal cooperation' (implemented)	
			Planning Association (Planungsverband) (newly created)	Council of the Region (Rat der Region), and intermunicipal cooperation
Representation, composition	• compulsory membership for 43 local authorities • directly elected regional representation • was integral part of institutionalized local government • main weaknesses: 1 covered just part of functional urban region 2 lacked executive powers vis-a-vis local authorities	• formal tier of subnational government at county level, replacing counties and upper tier functions of unitary urban authorities (i.e. cities lose status) • legitimated through directly elected regional parliament • encompasses whole functional region	• 75 representatives (councillors) of all municipalities in the conurbation, cities have greater voices (Frankfurt counts for 12) • no direct democratic legitimacy • mere 'talking shop'? • policy implementation through local authorities	• two reps for each municipality with over 50,000 inhabitants (mayors and county leaders (Landräte)) • indirect legitimacy • task specific aggregation of municipalities ('marriages of convenience')
Established	1970s	1996 (proposed, not implemented)	2001	2001
Tasks, responsibilities	• far-reaching responsibilities for city-regional land-use planning and including waste management	• planning and development • economic development • education • company status for waste management	• regional planning	• council to ordinate cooperation between municipalities • collaboration ad hoc based on 'problems' • powers limited to specific tasks

Source: After Bördlein (2000) and Sturm (2001).

functional interrelationships. So, it was not surprising that the Association was abolished at the end of 2001 and replaced by, as it turned out in the end, a much weaker, voluntary, collaborative arrangement. Two successor models have been promoted, one by the social-democratic SPD *Land* government of Hesse, and then, after political changes, one by the conservative CDU. The former proposed essentially a new regional tier of administration and planning in replacement of the 'counties' (*Kreise*), and including unitary municipalities. The name 'regional county' (*Regionalkreis*) reflected its intended firm position in the administrative hierarchy. The latter proposal, by contrast, was a much more watered down version without any statutorily backed position for a new regional authority. Instead, the new arrangement introduced after the *Land* elections in 1999 provides a tripartite regionalization with very weak institutionalization. There is one Regional Planning Association for developing a regional strategy plan which local authorities have to join and co-finance. The association covers a wider territory than the regional organization it replaces. Instead of 43 there are now 75 municipalities (Sturm, 2000). Its responsibilities are limited to regional planning and marketing. In addition, there are provisions for task-specific, smaller groupings of local authorities into subregions. These 'policy regions' may exist for one specific project only, and several such subregions may overlap both territorially and in their membership, that is one municipality may participate in several policy regions. To provide some control over so much potentially competing planning and policy making, a Council of the Region, consisting of the leaders of the participating municipalities, is to hold the various strings together. Lacking any meaningful power, critics have dismissed it already as merely a talking shop (Bördlein, 2000). Table 4.3 contrasts some of the key features of the two models of regionalization in the Frankfurt area.

Regions and regionalization in federal (devolved) states – summary

The cases in this section have confirmed the inherent competition between local and regional interests when it comes to the need for collaboration between municipalities. The stronger statutory position of subnational government in a federal rather than unitary government system seems to have somewhat exacerbated the competitiveness of interests. While, on one hand, regions have gained a 'naturally' stronger, recognized position under federalism, on the other there seems a stronger localist tendency, fuelled by a perceived considerable policy-making autonomy. This may, ultimately, undermine regional government at the city-regional scale, as the German case studies revealed. Strong, competitive localism undermined seemingly well established city-regional planning bodies and 'repatriated' almost all powers and responsibilities transferred to them earlier. The effect has been a distinct weakening of governance at the city-regional scale, which is in danger of being reduced to little more than a vaguely coordinated accumulation of local policies. Compared with unitary states, city-regional governance seems to be facing disempowerment through competitive localism, whereas in a unitary setting it is central government which is likely to feel that effect. Much of this has to do with the fact that city regions as a scale of policy making seem to be falling between the two stools of traditional regional scale governance on one hand, and well-established, confident local government as autonomous policy makers, on the other. Under federalism it seems to be in the latter's self-interest to collaborate and engage with the region. Polycentric city regions are, in principle, more likely to face competitive localism under a federal arrangement, whereas a dominant major city may exercise sufficient regionalizing pressure, if only for its own benefit. There is a danger, however, that, depending on the size of the other communities in such a region, resentment and, ultimately, non-cooperation may ensue. Before moving on to a more detailed analysis of individual cases in city-regional governance in federal and unitary contexts, the following sections explore the special circumstance of regionalization in national border areas, and evidence of 'new regionalism' there.

International regionalization and European policies

Nation-state borders are often particularly significant in shaping regionalization but their role in this process has been reinterpreted in recent analyses. Earlier studies of borders and border regions, based on trade flows, tended to stress the dividing effects of borders and the territorial separateness of border regions (House, 1981). More recently, the changing nature of borders in Europe and the growth of political initiatives, specifically for border regions, have shifted the focus considerably to the potential role of border regions as vehicles for European integration. These changes involve a fundamental change to the perceived nature of a border from 'separating' to 'bridging'. This includes the participation of border regions in the re-territorialization and rescaling of governance in Europe (Anderson, 1996; Cox, 1997). At the same time, doubts have been expressed about the reality of such transborder regionalization (Perkmann, 1999) against a backdrop of now nearly 120 examples of cross-border networking in Europe funded by the INTERREG, PHARE and TACIS programmes (AEBR, 1996). Scott *et al.* (1996) suggest that, collectively, they amount to an informal policy for border regions, both within the EU and between the EU and Central European countries.

Despite claims about a decreasing importance of international borders (Ohmae, 1995), considerable research in Europe has identified important and distinctive territorial, and governmental, consequences emerging from current patterns of regionalization in border areas in Europe. These include implications for national governance (Jessop, 1995; Duchacek *et al.*, 1988), or cross-border urban cooperation (Church and Reid, 1998, 1999; Krätke, 1999), or the role of historical, ethnic and cultural communities that span international borders (Perkmann, 1999), or general economic development potentials (IAW, 1992). The differing accounts all suggest that rather than separating, border regions are essentially a part of the wider political and cultural dimensions of regionalization in Europe, as well as part of traditional policies of counteracting the effects of peripherality. There is evidence that EU-funded, incentivized cross-border initiatives do have a complex influence on the institutionalization of these regions (Church and Reid, 1996; Perkmann, 1999), including those more particular cases along the former Iron Curtain. The INTERREG programme, established in 1989 to facilitate cross-border cooperation between 'East' and 'West', reflects the emergence of two scales and natures of transborder regions. Initially focused entirely on the immediate border areas along the former Iron Curtain, the programme has increasingly been expanded to also address wider, less formalized transnational regions, thus adopting effectively the traditional wider scale regional policy approach of the Commission. INTERREG initiatives occur in two main spatial scales: (1) transborder cooperation between neighbouring regions separated by an international border, and (2) since 1996, much wider transnational cooperation between the EU and neighbouring Central Europe. Seven such transnational mega regions have been established so far (see Eltges, 1997: 379–386).

There are considerable 'incentives' within the EU for transborder cooperation between regions. The INTERREG programme amounting to ECU 415 million (Ahlke, 1997). These respond to the challenges of globalization and the opening up to eastern Europe. The new pressures transcend national territories and require more far-reaching border-crossing policies. Initiatives revolve primarily around planning policies and strategies at the regional level (INTERREG IIC). The scale of 'region' has been increased to exceed cooperation primarily between localities in the immediate border areas, to much wider interregional linkages, including large transnational regions sweeping across Europe (Map 3.3), such as the Union of Baltic Cities. This might be viewed as a shift from 'old style', neo-classical *interregional government* to 'new regionalist' *transregional governance based on networks*. The nature of 'territory', however, may vary between city network-based regions bounded by intraregional functional relationships, and centrally established regions as designated entities of common policy agendas for anyone who is part of that region. The formal territoriality of the Euroregions suggests a continuous emphasis on the former as the foundation of real, financially supported, cross-border regionalization (Anderson and Hamilton, 1999). The large transnational, 'super regions' (see Map 3.3) include eastern, central and southeastern Europe in anticipation of future expansions of the EU and its policy-making arrangements. Depending on their geographic location, some countries are part of more than one super region. Germany, for instance, belongs to four such transeuropean super regions (see Map 3.3), reflecting its geographically central location in the EU, as well as the regions' less formalized, network-based nature.

Conventional, territorially bounded cross-border regionalization

Border regions are explicitly acknowledged as part of the EU's structural policies, because of their inherent structural economic weaknesses. In the early 1990s, 39 out of 76 economically less well performing regions were border regions. Yet, at the same time, about one third (32 out of 91) of economically stronger regions are also border regions (see Brown and Cäniels, 1997). This implies that it is not necessarily the border status per se that holds those areas' development back, but that other factors may be at least as, if not more, important. This would suggest a need for equally diverse policies and a more detailed response to the variations in economic circumstances within and between border regions.

The regions along the former Iron Curtain occupy a uniquely important geographic-political location at the meeting point between the EU and the aspiring future member states of post-socialist Central Europe. Socialist, *dirigiste* planning treated borders as dividing lines with strong defences against a neighbour portrayed as hostile. Border controls were perceived as important opportunities to parade and visibly assert national sovereignty in a Soviet directed and controlled Eastern Bloc. Borders were thus distinctly divisive in character and cut through existing cultural and historic territorial relationships. The shift from politically motivated division towards integration thus marks a particularly strong paradigmatic shift in the projected role and understanding of border regions both from within and without the area. By contrast, borders within the EU have increasingly become less visible as an expression of state territorial identity. The Schengen agreement of 1998 can be seen as the culmination of a politically driven annihilation of borders as dividing lines. Nevertheless, as anyone who has travelled through the German–Dutch–Belgian border areas, e.g. near Maastricht, will have noticed, there continues to be succinct evidence that territorial separateness remains, if formally no longer enforced: upon crossing the 'invisible' border all car licence plates in the streets change from one nationality to another, rather than showing an international mix.

The smaller, formally declared (by the EU) transborder Euroregions (or Euregios) (Map 4.3) define a group of local authorities along a border, which are expected and encouraged to engage in collaborative cross-border initiatives (Scott, 1998). The emphasis is on institutionalizing a cross-border mechanism for the administrative process of transborder development planning. Mutual consultation is intended to build bridges for channels

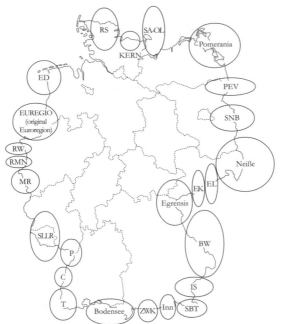

Map 4.3 Euroregions in Germany.

Source: Information from BBR (2000): Raumordnungsbericht 2000.

of communication and, ultimately, collaboration. Informal arrangements between non-governmental actors, including universities and other institutions, are intended to broaden and reinforce the bridge building. The three countries of Benelux and the Dutch–German border offer a model of institutionalized, multilevel cooperation (vertical and horizontal). It is here that the concept of the Euroregion originated, based around intergovernmental planning commissions as institutional joint context (Miosga, 1999). The focus on planning as the main concern of cooperation and their role as a representational platform of high-level government agencies and other organizations implies a technocratic focus of such initiatives. Mechanisms for popular input into these processes provide for some democratic legitimacy. 'Cross-border regionalism manifests itself not only as systems of governance, but also as interests and development priorities articulated in the form of strategies that guide co-operative action' (Scott, 1998: 609).

The objectives vary somewhat between the EU's inner and outer Euregios: the internal, longer existing ones, focus on establishing and, in many cases, furthering, existing networks of communication and cooperation between and within levels of government across the borders. This includes real measures like integrated cross-border public transport. In contrast, the Euregios on the outside of the EU, primarily along the former Iron Curtain, emphasize basic physical development projects, such as road links (Scott, 1998), to physically reconnect the previously ruthlessly separated regions, and put the fundamental preconditions for economic development in place. The initiatives cannot, however, alleviate the inherent differences between urban and rural (peripheral) localities *within* the regions, and pointing to the problems of structural heterogeneity for region building. Referring to the German–Polish border, Krätke (1998) identifies three basic types of cross-border cooperation, based on the difference between the scale of links at transnational and interregional links, where there are also further scalar subdivisions: (1) long-distance, international transborder links, (2) supra-regional links from metropolitan areas on one side of an international border to the (less urban) region on other side, and (3) regionally integrated links *between* border regions at a more local level.

So far, institutional, cultural and operational differences have made it difficult to create genuinely collaborative transborder regionalism. Formalization does not simply overcome these problems in a technocratic way. Other factors matter, like the internal organization of the institutions and their practices. The German–Polish border regions, for instance, face considerable difficulties in establishing integrated policies. Much of this is owing to a lack of experience in utilizing the available structural administrative framework effectively. Established legacies of of government practices, such as centralist versus federal traditions and experiences (Scott, 1997), are important for the institutional-political efficacy in using available resources and opportunities. Krätke (1996) therefore refers to the importance of creating a supportive cross-regional 'milieu' as an essential preparatory step for the successful operation of Euroregions. This may be in the shape of loose, network-based collaborations based on common purpose, or fixed, formal institutional structures acting as facilitators or implementors of cross-border collaboration and policy. The national differences in subnational governance have contributed to, at times, a somewhat haphazard and sporadic form of collaboration and communication, often to chase available funding, rather than reflecting on a sense of genuinely shared policy goals. As a result, there is little vertical integration between administrative tiers across borders. Any communication remains strictly at the same government level on both sides and follows the respective governmental hierarchies. This is likely to be slow and cumbersome. One question is whether common economic interests, or the availability of funding, are sufficient as the main driving force behind cross border cooperation to create a sustained sense of common interest in policy making towards establishing desired growth regimes. Or are historical and cultural factors so important as to influence the priorities and interests of the various actors within the border regions?

Network-based transnational super regions in Europe

The more recent type of cross-border regionalism is the large-scale transnational regions sweeping across Europe, held together by mainly informal networks, and showing a distinct absence of clearly defined territoriality. Much of their purpose is about external visibility as a growth region or 'core region' in Europe, and thus a positive image for any territory associated with, or part of, it. Not surprisingly, therefore, a lot of bargaining and politicking has been going on between cities, regions and states on the one side, and the European Commission on the other, to press for inclusion in one of the 'super regions'. Much of Europe has thus been covered by variously themed super regions, mainly based on urban clusters and clearly visible geographic features.

Each of the transnational super regions has its own developmental strategies and priorities, implying region-specific agendas and thus an inherent sense of common purpose as their foundation. The basis of a genuine sense of regional belonging varies considerably, reflecting differing aspirations, legacies and expectations and thus the degree to which these regions are realistic. In some cases this may lead to failure. Thus, in central southeast Europe the Carpathia super region was established in 1993 between Poland, Slovakia, Ukraine and Hungary. Since then, Slovakia has left the arrangement to pursue its own agenda, suggesting a rather low effectiveness of such politically driven token regionalization vis-à-vis a strong national

Table 4.4 INTERREG IIC super regions of cooperation in the EU

Region	Participating countries
North-west European Metropolitan Area	Belgium, Netherlands, Germany, United Kingdom, Ireland, France, Luxembourg
Atlantic Arc Region	United Kingdom, Ireland, France, Spain, Portugal
Mediterranean Region	France, Italy, Spain
South-west Europe Region	France, Spain, Portugal
Central south-east Europe Region	Germany, Italy, Denmark, Spain, United Kingdom
North Sea Region	Netherlands, Germany, Denmark, Sweden, United Kingdom
Baltic Region	Germany, Denmark, Sweden, Finland

self-interest. The Euroregions along the Polish, Czech and German borders with their shared cultural-historic legacies are much more successful by comparison.

At the wider transnational scale, cities and their regions are the common feature holding together the North-West European Metropolitan Area with its distinct urban focus and emphasis on sustainable development. However, this region may effectively be little more than an amalgamation of individual local economic and policy-making entities with their separate identities. Functionally, or historic-culturally driven collaboration establishes city network-based subregions, such as the Paris Basin, interurban cooperation across borders (Nord-Pas de Calais, Alsace) and the port cities along the Channel (Church and Reid, 1998). An urban focus based around a network of medium-sized cities, intercity competitiveness, and urban growth (suburbanization) have emerged as key ingredients in transborder and especially transnational regionalization (Robert, 1997: 419–422). The comparatively successful Baltic Region is a case in point. Held together by common geography and historical and economic affinities, it depends on the development of strong urban areas as growth centres of international relevance. The urban centres, referred to as 'pearls', are connected by growth corridors ('strings'), outside of which more peripheral areas exist as 'patches'. There is thus a clear admission that this transnational regionalization encompasses very different subregional entities with considerable variations in development potentials and thus policy requirements, actor networks and established territorial identities. Much emphasis is placed on establishing a functional, institutional structure to run the programme (agreed by ministers of the participating countries) to give this super region a sense of reality and greater visibility. The East–West INTERREG programmes, for instance, facilitated the Union of the Baltic Cities as an international city network aimed at maximizing the economic potential within the region and selling the region to the outside world. Placing its seat in Gdansk, Poland, is intended to highlight the weakening of the East–West border and emphasize the historic economic and cultural commonalities throughout the Baltic region (see also Baldersheim and Stahlberg, 1999).

Mono- and polycentric regionalization in unitary and federal states: summary

The review in this chapter of the nature of regions and regional policy across the European Union has revealed interesting differences and similarities between countries. The comparison looked at two fundamental factors circumscribing forms and processes of regionalization: the type of state organization, varying between federal and unitary, and the internal structure of regions as circumscribed by the size and number of cities. The former factors circumscribe the position of regions both in terms of their recognition as entities of territorial governance in their own right and their scope for autonomous planning and policy making. Unitary systems showed an inherent ambivalence about regionalization, drawn between the recognition that regions had become a sine qua non in a globalizing economy and subsequent territorial competitiveness on the one hand, and uneasinesss about a likely loss of control and increased bureaucracy, on the other. As a result, they seek to retain as much control as possible by limiting regionalization to narrowly defined areas of responsibility, primarily aspects of economic development and place marketing. For the same reason, unitary states tend to favour non-territorial regionalization, and informal, network-based arrangements which can be changed (and removed) more easily. Informal, network- rather than territory-based regions are seen as less likely to develop their own political power base which might pose a challenge to the central state or, indeed, established local autonomy. Networks may seem easier to manipulate than firmly established territorial government. The internal structure of regions adds to the path of regionalization. Monocentric regions appear more likely to develop a clearer political objective and policy direction than polycentric regions with their many players and diverse, competing interests. The former are thus a potentially greater challenge to the unitary state than the latter, because intercommunal competition for regional influence is more likely to allow, or even require, a 'guiding' involvement by the central state. *Within* regions, a monocentric structure seems more likely to accept a more formal approach to regionalization than poly-centric arrangements, because the main city will expect to maintain a controlling influence, while mutual distrust and localist considerations in a polycentric region are much less likely to accept additional scales of territorial government and may respond with non-cooperation, undermining any regional agenda.

These concerns and anxieties appear in a somewhat different light in a federal state structure. There, multi-tier territorial government with clearly allocated responsibilities is a fact of political life. Regions are therefore an integral and accepted scale of government and as such possess political-institutional credibility and visibility. The main challenges and potential difficulties here are anxieties about possible shifts in the power distribution between the existing tiers of government. The concern is that changes, once initiated, can easily become permanent features of the federal structure. Protecting established self-interest is a very powerful force. This includes both the government levels above and below the 'region'. Central governments seek to maintain their powers, just as local governments, thus preferring the status quo in institutional terms. New forms of regionalization are thus preferably informal, voluntary and time-limited arrangements which are less likely to pose a challenge to any existing power distribution and allow withdrawl from any arrangements in the interest of local (or central) concerns. Discussions on regionalization tend to be linked automatically to territoriality and government hierarchy, because these are the backbone of federal state structures. The focus on structure, hierarchy and territory makes any modifications appear as potential challenges to the existing order. In this way, federal state structures, while giving regions a clearly recognized role in governance, are more likely to be resistant to changes and varying forms of regions than unitary states, where new forms of regionalization can simply be introduced by the central state without significant challenges by the lower levels of government.

5 Formal regions and regional governance in England and Germany

Centralized and decentralized regionalism

In the next three chapters the discussion moves on to the detail of regional governance and planning in England and Germany as two examples of, respectively, a highly centralized and a decentralized organization of government. The review of European trends and cases in Chapters 3 and 4 emphasized the variation in regional experiences. Chapter 2 suggested that theoretical debates about regionalism were also moving in directions that could accommodate and attempt to explain the evident institutional and policy variation on the ground. In particular, we have identified mono-centric and polycentric regions as offering very different bases for the development of regional governance under different national frameworks of government organization. In Chapters 6 and 7, the nature and management of regional governance and planning issues will be examined in detail for these differing types of region in both countries. In this chapter, the objective is to trace the formal arrangements of regional government in the two, contrasting, constitutional circumstances of England and Germany. Formal powers and responsibilities set the conditions for new developments in regional governance, be they encouraging or discouraging collaboration across, and within, levels of governance. In particular, the focus will be on evidence of a possible positive relationship between the degree of centralism of the state structure, interlocality collaboration and the degree to which regions are created from the top down or especially the bottom up. In reality, regional governance may not be as clear cut. Approaches to regions and especially regional policy are much more individualized and dependent on particular local circumstances and institutional capacities. Furthermore, recent institutional developments in relation to spatial planning and economic development will be examined, e.g. the creation of Regional Development Agencies in England and Regional Development Concepts in Germany, and their impact on regionalization. It is this relationship between institutions which will shape, and be shaped by, the position of regions as policy-making entities.

Constitutional arrangements for regions as territorial units of government circumscribe their scale and competence as separate and identifiable entities of policy making. By their very nature, regions are sandwiched between the local and national levels of government. In most countries these are the clearly identifiable institutions of government leaving regional concerns as a grey area of intergovernmental responsibility, or a relatively new intermediate level. National governments manage through subdividing national territory and developing more detailed nationally defined policies for each area which functions essentially as a territorial container of national policies. At the local level, there is concern about, and resistance to, perceived infringements upon their powers and responsibilities. More profound constitutional settlements periodically change the scales of government through devolution and establishing new intermediate tiers. England and Germany illustrate two very different sets of constitutional arrangements for regionalization a strongly centralized system and traditionally weak regionalism, and a federally arranged system with a strong regional scale.

Previous chapters identified the often complex interactions of governmental and non-governmental forces which shape regionalization. In our examination of regional development in England and Germany several sets of factors come into play:

1   The position of regions in the institutional state hierarchy and division of policy responsibilities between central, regional and local levels of government and thus tendencies towards centralized or decentralized regionalism. This also sets the scene for the involvement of a variety of actors from within and without government.
2   The degree of democratic legitimacy of regional policy through region ally operating local or central government, circumscribing the credibility and impact of either.
3   The understanding of what constitutes relevant regionalization: for example, static technocratic planning or more dynamic policy making.
4   The awareness of new forms of region building and regionalization in policy making.

In Germany, the explicit general statutory requirement for local authorities to advance the welfare of the local population implicitly includes regional initiatives. It is through the impact on local circumstances that regional initiatives obtain their legitimation. In both countries the legitimacy of local government is based on the representation of a local electorate through an elected multi-purpose government body, the local council or *Stadt-Gemeinderat* respectively. In the absence, for the present, of regional parliaments in England and no, or no direct, democratic representation between local and *Land* level in Germany, it is the local or central governments that act as (indirect) legitimators of regional initiatives.

The scope for taking initiatives, however, varies between the local governments in the two countries. In England, the emphasis is on providing statutorily defined services, including those of wider regional relevance. This contrasts with the far-reaching local autonomy granted by the *Grundgesetz* (Basic Law) in 1949 as the foundation of the German Federal Republic's government system. This allows local authorities to take any measure necessary to advance local welfare. In addition, the federal system requires a constant negotiation process between the various tiers of government, with continuous compromises necessary about the statutorily less clearly defined nature of the region.

The stronger top-down tradition of policy making in England is less amenable to such negotiation and challenging of regional responsibilities. This obviously affects the engagement with EU regional policies and initiatives. In England all formal communication, especially that involving funding, goes via the national government whereas in Germany such communication is decentralized to the *Länder* level, although this may very well act as a central state. This adds a certain degree of regionalism and regional variation in political lobbying

Scope and provision for locally-led regionalization in England

During the 1990s, political debate in England increasingly acknowledged the role for regional governance as an effective means for marketing regions as a more promising way of attracting inward investment. National objectives coincided to some extent with the increasing interest of local governments in marketing and development at regional scale. Local governments, especially the larger urban authorities, developed their own regional agendas and policies, including forming cooperative networks outside the centrally controlled sphere of formal local government action. Such informal arrangements are not unique to England, as illustrated in the previous chapter, but there are considerable differences in the way in which local government has the scope to develop and implement its own regional agenda.

The status of subnational government varies considerably between England and Germany, and with it the freedom to engage in local–regional policy making, including cooperation to define regional development and policy objectives. There are important differences in the standing of central government and its involvement with regional matters vis-à-vis local government. Directly related to this, the position of local government in representing its interests in competition with the central state also differs considerably, and this includes the involvement with regional matters. In England, in contrast to Germany where the central government function rests with the 16 *Länder*, this role resides at the national level.

It is only since 1999 that some form of large-scale regionalization has occurred in the UK with the devolution of powers to Scotland and Wales (Figure 5.1). The difference in spatial scale is important, because it affects the extent to which regional, and local, matters and interests influence central government policy considerations. Such spatially more varied and hence

BRITAIN	GERMANY
CENTRAL GOVERNMENT SPHERE	
NATIONAL GOVERNMENT executes central control over local government.	**FEDERAL GOVERNMENT (NATIONAL)** no direct link with local government.
COMPONENTS OF THE UK Devolution of powers to Scotland, Wales and Northern Ireland with own parliament/ assemblies. England without separate representation	***LÄNDER* (STATE) GOVERNMENTS (16)** execute supervisory role as central government for municipalities (*Kommunen*).
REGION: SPHERE OF COMPETITIVE INFLUENCE BY CENTRAL AND LOCAL GOVERNMENT	
A. Central Government-Led Regions	
Government Offices as joint regional representation of national government departments, established to enhance communication with local govt., primarily for more effective implementation of central policies. **Regional Development Agencies** as government appointed quangos to promote regional economies.	***Regierungsbezirke*** (Administrative regions): regional representation of *Land* governments (varies between *Länder*) to control and supervise legality of local administration, and facilitate central *(Land)* government policies. They also help to convey local governments up to the centre (similar role to GOs in England). **Planning Regions:** established by *Land* government to implement regional planning (local definition in some *Länder*).
B. Local Government-Led Regions	
A. Regional Assemblies, Regional Chambers as (democratic) regional representations. **B. Shire Counties** (non-metropolitan county councils): statutory body with democratic representative councils. **C. Unitary Metropolitan Authorities** (since 1986), combine local and county functions (similar to unitary urban authorities in Germany).	**A. Planning Regions,** locally defined by group or municipalities. **B. *Kreise*** (non-metropolitan). Statutory bodies as group of municipalities, legitmated through local councillors delegated to assemblies (similar to regional assemblies in England). *Kreise* provide local service of higher centrality for a group of municipalities. **C. *Kreisfreie Städte*** (unitary urban authorities): Kreis function combined with local function in larger cities (here focus on subregional *Kreis* functions).
LOCAL GOVERNMENT SPHERE	
A. Metropolitan Districts (incl. London Boroughs). **B. District Councils** (rural communities and small towns) incl. subdivision into 'neighbourhood'-based representational bodies of parish councils.	**A. *Kreisfreie Städte*** (here: local function) (see above). Compulsory creation of neighbourhood-based representation *(Bezirksvertretungen)* with no governmental function. **B. *Kommunen*** (rural communities and small towns).

Figure 5.1 Regions in the government hierarchics of Britain and Germany.

implicitly more area-specific government responsiveness was the main consideration behind, first, devolution in the UK and then, the shift towards regionalization in England. As part of this process, large local authorities, such as the main metropolitan areas, will regain political presence and influence at the regional level and even beyond. But the large urban authorities present a political problem for national government, as evidenced in the abolition of the metropolitan county councils by the Thatcher government in 1986 (Cochrane, 1993) and, more recently, the political infighting around the election of a London mayor and wider resurrection of demands for a more formal city region status.

Until the devolution reforms of the late 1990s, the UK possessed a strongly bipolar government system with a relatively weak local level and strong central tier. This has encouraged two tiers of policy making, where the centre can challenge and modify its degree of direct involvement with local and regional government. Describing English local government as a 'creature of Parliament' (Hampton, 1987) highlights its subordinate position and relative weakness within the confines of *ultra vires*. 'The term *ultra vires* means "beyond the powers"', where 'a statutory corporation can do only those things which it is authorised to do by statute, directly or by implication. If such a corporation acts otherwise than in this way its acts are *ultra vires*' (Cross, 1981: 4). The position of local government may change as defined and redefined by Parliament, in response to changing perceptions of the role of local government, and this includes local engagement with regionalization.

National government may change territorial structures and install or remove actors and organizations engaged with regional matters. The institutionalized government system is thus strongly biased in favour of top-down, centralized regional governance, leaving much less scope for locally defined policy making. Local government is shared by counties, districts and central government agencies. Formally, the regional dimension is part of local government in the shape of the counties (*c.* 100,000 to 1,500,000 population) where they operate as the upper tier of local government. The lower tier, with its specifically local focus, is made up of the districts (*c.* 20,000 to 1,000,000 population). Elsewhere, unitary authorities take the strategic role. A similarly integrated position for larger urban areas can also be found in Germany as the *kreisfreie Städte* (literally, 'county-independent cities'). This arrangement acknowledges the different circumstances for urban and rural localities and many local political wrangles. In London, as we shall see, the responsibility for planning decision making is shifting between levels of government in response to political consideration by the centre.

Structures of local government have been changing, but so have the internal arrangements of central government, most notably in the establishment of the Government Offices for the Regions (GOR) in 1994. This deconcentration of central government departments finds some parallels in

Germany in the *Regierungsbezirke* (regional administrative units for a *Land* government, each with a head office). The similarities to the Government Offices in England, include especially their ambivalent position between government levels: acting as central agents in the region and implementing 'top-down' policies, while also acting as 'eyes and ears' of the centre to convey local concerns, interests and lobbying up the governmental hierarchy. In Germany, this two-way flow of political bargaining and lobbying is called *Gegenstromprinzip* (principle of counter-flow [of political pressures]) as part of 'cooperative federalism'. This is of particular importance for regional economic policy making between central and local interests because the quality of such exchanges will influence the nature of policy responses to economic regionalization processes.

In 1999, there was a further addition to the institutional landscape with the introduction of Regional Development Agencies (RDAs) which can be seen as a step towards decentralization of regional governance. To what extent the RDAs can effectively conduct genuinely region-based policies vis-à-vis the established structures and practices of centrally directed region-alization through the Government Offices, will be examined in the two English case studies in the following chapters. Unique for London, at least so far, is the establishment of an elected mayor and assembly responsible for strategic planning and policies throughout Greater London. This explicit city-regional scale of governance, and its institutionalization, will be explored in detail in Chapter 6.

Planning for regions and cities

The boundaries of official, administrative regions in England have changed from time to time since the enthusiasm for national and regional planning in the early 1960s. Government interest in regions has always concentrated on economic management and planning. At the end of the 1960s, however, the focus shifted from the larger regional scale to the level of newly defined metropolitan regions around the big cities, again, for reasons of growth management. This tier of regional government was subsequently abolished in 1986 and since then there has been no formal tier of government between individual local authorities and the centre.

These changes in formal structures reflect both a historic distrust of lower level government by the centre and particular party political projects. The Conservative governments of the 1980s and 1990s installed new types of economic development and city regeneration bodies, bypassing and hence weakening the remaining level of elected local government. Training and Enterprise Councils were established to manage subregional training programmes (responsibilities subsequently passed to Learning and Skills Councils) within their own separate territorial boundaries. In the 1980s, Urban Development Corporations managed large-scale redevelopment, with local government excluded from decision making and reduced to little more than bystanders (Imrie and Thomas, 1999). These local agencies themselves were of limited life, and their responsibilities for managing investment in land and property were passed on to the national level English Partnerships from 1994, thus effectively centralizing land- and property-based regeneration. The operational programmes of English Partnerships supported EC Structural Fund investments in Objective 1, 2 and 5b areas, as well as national urban regeneration policies. Regional planning has had some institutional continuity in joint local government planning boards but few staff were assigned to regional planning issues. Formal regional planning in Regional Planning Guidance is produced by national government. England is thus characterized by centralization, multiplication of development agencies and, since the 1980s, a much more visible role for business leaders in city and regional governance. National policy initiatives, through Development Corporations and other economic development tools clearly emphasize localized, especially urban, perspectives, to the detriment of wider regional thinking.

Planning at different regional scales finds itself in difficult terrain between central and local governments. At the scale of the old Standard English Regions (see Map 5.1) plans are a central government responsibility and only very recently are there signs that local and regional factors may begin to influence Regional Planning Guidance (RPG). The prospect of elected Regional Assemblies, and the need to coordinate plans with the economic strategies of RDAs, offer potential scope for more bottom-up influence. RPG is also under pressure from above; there is a need to take notice of European policy and make explicit reference to the ESDP, not least for financial reasons. Regional strategy is also subject to a new regime of sustainability appraisal (DETR, 2000). The impact of such close guidance and scrutiny of regional plans is discussed in Chapters 6 and 7. At city-regional scale planning has a chequered history. Abolition of metropolitan governments in the 1980s was justified in part as a response to the failure of strategic planning at this scale. In London, the Greater London Development Plan had taken years to produce amid extensive controversy in the later 1960s and early 1970s. By the early 1980s, a replacement plan was vetoed by central government. Political controversy undermined whatever value such plans may have had. In the climate of the 1980s, strategic planning was not compatible with a market driven approach to (speculative property led) development of the cities and to state intervention through development corporations with their own city building agendas. Abolition of the metropolitan scale meant both localization and centralization at the same time through a shift of strategic planning to local governments under the guidance and control of the centre.

The need to coordinate this scale of metropolitan plans was the task of a system of Strategic Planning Guidance. This SPG could be brief, reflecting Thatcherite ideology with emphasis on minimal state intervention and maximum scope for the private sector. The Merseyside Guidance issued in

(a) Old UK regions (pre-1994)

(b) New UK regions since 1994

Map 5.1 Old and new UK regions before and after 1994.

the late 1980s, for instance, ran to just six pages, supporting private sector investment and the local development corporation. The views of local governments on economic development and housing policies were generally ignored, but local governments' Unitary Development Plans had to follow this limited but clear guidance. This minimalist approach was, however, not transferred to the new system of Regional Planning Guidance in the early 1990s, when a more conciliatory tone was adopted by John Major's government towards the local level. Local authorities were asked to form Regional Planning Conferences to offer advice to central government on the scope of RPG. By the mid-1990s there was national coverage of RPG. The first round of SPG was never revised and RPG took on the strategic role instead.

There were two significant changes in the early 1990s. Both can be seen, in differing ways, as responses to the increasing influence of European regional policy and sources of funding. The Government Offices for the Regions gave central government a revived regional structure based largely on old boundaries. The Government Offices aimed to coordinate government functions – planning, inward investment, training and transport. In fact, many core budgets were not devolved from the centre, and the GORs managed only a small part of state expenditure in the regions. Control over European bids and budgets was retained at the centre. The GORs were encouraged to develop better central–local relations, which can be seen as a limited attempt at improving regional coordination of central programmes. Local authorities were sceptical of the new regional administration and, following a decade of aggravated central–local conflict in development policies, such mistrust (or even cynicism) does not come as a surprise. The first change in the 1990s was therefore this increased, if limited, regionalization of central state machinery. There is evidence that the GORs have moved towards a more differentiated, region-specific approach, expanding their initial role of 'strengthening (of) central government's grip on regional structural fund networks' (Bache, 2000). Within the overall centralizing structure, different approaches to partnership, reflecting the varying degree of homogeneity and commonality in interests among actors, can be detected in the regions. The East Midlands region, for instance, emerges as relatively fragmented and less innovative in policy making. Neighbouring Yorkshire and the Humber region, by contrast, 'has adapted to the partnership culture more quickly than its East Midlands counterpart' (Bache, 1999: 20), resulting in greater policy-making flexibility. These differences are also evident from their Regional Economic Strategies which mirror quite varied involvement and collaboration of actors, and contain different degrees of detail (Benneworth, 2000).

The second important change in the 1990s was evidence of de facto regionalization through various bottom-up initiatives. Largely in response to the perceived centralization of Conservative central governments, local authorities began to cooperate more positively in informal regional structures *inter alia* to strengthen their lobbying position. Regional Associations of local authorities operated as informal arrangements between local authorities in the nine English regions, aiming at including regional consideration into local planning, albeit to varying extents (Mawson, 1997b). New arrangements for cooperation included the creation of regional identity through more promotional and lobbying objectives with a clear emphasis on marketing to attract new investment. Regional identity was also seen to be an important part of successful lobbying for European funds. The boundaries of areas eligible for funding, however, also necessitated intraregional, locally-based cooperation, albeit with a pragmatic focus on obtaining EU funding rather than pursuing policy-making alliances. The overall impact of EU funding on government are not fixed and may encourage more variable allegiances between localities. Both central and local levels of government can be argued to have gained influence through the process (see Martin,

1999; Bomberg and Peterson, 1998). However, regional associations of local government, led by the big cities, formed a strong lobby within the Labour Party, and in the mid-1990s regional reform became part of the political agenda. The stronger regional agencies operating in Scotland and Wales gave the English regions a model to work on.

This pressure from Labour local government can be argued to have pushed the new Labour government, in 1997, towards administrative reform geared to economic development. New Regional Development Agencies, operational from 1999, were to 'build up the voice of the region' (DETR, 1997), giving a 'sharper regional focus' (DETR, 1997: foreword). Nevertheless, there continues to be a strong centralist undertone: The RDAs based on the 12 regions (Map 5.1b) have centrally determined constitutions and budgets. Their executive boards are appointed by government primarily from the business sector to underline the focus on economic development, but include civic leaders and university and trade unions representation as a reference to democratic (regional) legitimacy. The primary role of RDAs is to promote competitiveness, innovation and investment, and develop comprehensive regional development strategies. But there is only limited control of resources, and thus effectively with rather limited *actual* power. Some existing conventional regional development programmes, for example Regional Selective Assistance to companies, remain under central control (the bulk of these funds, however, support projects in Scotland and Wales). The centre thus continues to run regional matters as a hands-on matter and makes sure 'that the primary line of accountability remains to Whitehall' (Anadyke-Danes *et al.*, 2001: 15).

The strong emphasis on competitiveness raised concerns about RDAs competing against each other. Regional Chambers made up of local authority and business leaders give some local legitimacy to the RDAs and may inject some impetus for a more localized approach to regional policy making. The RDAs and Chambers, however, have strong business representation which are potentially competing with government interests. This institutionalized emphasis on business interests is much stronger than in Germany with a more universal, government-centred approach to regional policy and planning. The overseeing role of the Chambers was initially unclear, given that the reporting lines of RDAs were to central government ministers. The reforms aimed for speed and to avoid lengthy debates about regional boundaries (GOYH, 1998). Existing administrative boundaries were, for the most part, maintained. The existing informal regional and subregional alliances, which had developed in the 1980s and 1990s, tended to be sceptical about the potential of RDAs and pointed to their limited budgets and unclear relations between RDAs and Chambers (DETR, 1998b).

The prominent role of the Chambers, and business representation within them, reflects the fact that institutional reform was about economic development rather than regional government per se. The legitimacy of the new bodies is sought at higher and lower levels. But at the same time, government is seeking to reform democratic structures in English cities. Those cities which adopt strong mayors might have a more influential voice at regional level. In London, for instance, reform was driven by the need to improve democratic accountability, yet also greater international visibility and competitiveness of London as a place, and the figure of a mayor plays an important role in that (see McNeill, 2001). The new regional boundary was that of the former metropolitan authority of the 1960s with the surrounding economic region split between two new RDAs (see Map 5.1b). In London, the RDA is under the control of the mayor, although with similar central budget controls as elsewhere.

However, the reforms of the late 1990s did not settle the debate about the balance of central, local and regional institutions of governance. The asymmetric devolution of powers in Britain gave the English regions model alternatives in Scotland and Wales with a genuine devolution of powers and accountability to the regional level. But public opinion suggested little enthusiasm for further devolution in England (Harding, 2000). The RDA, Regional Chamber and Assembly reforms also quickly appeared not to have fulfilled their promise. In the build-up to the 2001 election, the Minister for the Regions promoted the idea of further devolution to regional assemblies (*The Times*, 25 May 2000) and paving the way for elections to regional assemblies in those areas that wanted them. New government proposals on regionalism were published in early 2002. Contrasting with this renewed enthusiasm for devolution, however, was a review of the performance of the RDAs which drew conclusions in favour of centralization. The GORs and RDAs both aimed to draw together government programmes and policies in the regions, yet the Cabinet Office Performance and Innovation Unit's report identified the lack of coordination of government initiatives as a problem of intervention at local and regional level (Cabinet Office, 2000). The solution was seen as better coordination between Government Offices and RDAs, interministerial coordination, and overall control from the centre, not from DETR but from the Cabinet Office. Hence, only two years into the RDA experiment, the two forces of devolution and centralization continue to pull in different directions, and traditional tensions in the structures of English government have reasserted themselves. Initially following the 2001 general election, regional devolution seemed to have lost its place among Labour's priorities. The Deputy Leader of the party was no longer responsible for the central Department of Transport, Local Government and Regions, and the minister who had pushed through the RDA reforms had also moved to a new post.

Coordination and resourcing of the new institutions continues to present challenges. The Regional Chambers have few financial resources or staff, while the RDAs do have resources, but are constrained by their mission defined by the centre and subsequent detailed guidance. Thus, while the set of economic strategies produced by the RDAs offers an unparalleled review of the English regional economy, the strategies tend to follow national guidance rather than offering local perspectives. The RDAs cover large areas. Some RDA strategies identify specific subregional issues, and while working through subregional partnerships may offer greater legitimacy for the RDAs that may not have the resources to work effectively at both regional and subregional scale.

Draft economic strategies were vetted by central government. The RDAs' specific functions (and a large slice of their budgets) include administering the government's local area urban renewal programme. Whatever the analysis of regional problems, making priorities is circumscribed by centrally imposed roles. The termination of the national urban regeneration budget will, however, allow the RDAs to shift priorities. Still, the RDAs, charged as they are with regional competitiveness, and thus feeling under pressure to perform, may divert money into interregional competition for business investment. Competitive marketing is already a feature of the RDAs with, for example, English regions advertising their locational advantages in posters on the London Underground. There is a further problem of coordinating the range of strategies at the new regional scale. Economic strategies need to integrate with those for environmental planning and transport. The RDAs are not responsible for land-use planning and are thus deprived of an important tool of translating strategies into actual processes. The government's new arrangements for Regional Planning Guidance propose a style of spatial planning that would guide all regional strategies, thus continuing with a centralist approach. In Chapter 7 we examine the particular difficulties of such coordination in London for addressing regional variations. This continued emphasis of essentially *one* (centrally managed) regional policy towards the English regions may be seen as attempting 'to meet the challenges of the 1980s and 90s' rather than the more differential requirements in regional policy for the 2000s (Anadyke-Danes *et al.*, 2001: 28).

The RDA and Regional Chamber initiatives are in their early years. How they develop may depend as much on the perceived success of regional government in Scotland and Wales as on their own performance. The most important of the regional experiments may well turn out to be that in London. Leaders of other cities like the higher profile city-regional scale of the London reforms. At the same time, regional policies seem less 'new' than promoted Labour. RDA strategies have exposed subregional variations, but there is no matching system of regional accountability and policy making in place. London offers a model of democratic control of regional economic, transport and land-use planning for city regions through the Greater London Authority established in 1999 (HMSO, 1999), but even there, the ultimate control of policies continues to be held in Whitehall. The asymmetrical reforms of regional governance in England thus throw up the possibility, or even the necessity, of yet further reform.

Federalism and multilayered regionalization in Germany

Local and, by extension, regional, governments in Germany rest first and foremost on their constitutionally enshrined right of self-government as the basis of far-reaching autonomy in policy making. The main limit of this *Kommunale Selbstverwaltung* (municipal self-government), provided by Section 28 of the Basic Law, is the requirement that local policies and planning comply with the aims and guidelines set out by *Land* policies and planning, and thus clearly refers to the strictly hierarchical organization of government. At the same time, municipalities are provided with the statutory guarantee that they possess the far-reaching, general right to regulate all matters concerning the local community under their own responsibility, but within state law. This arrangement inevitably provides for arguments about where the line between necessary compliance with guidelines and local policy-making autonomy rests. The situation becomes even more complex by the fact that local government (*Kommune*) also acts as local agent of the national and *Land* governments for specified administrative tasks. It is not surprising therefore that over the last 20 or so years, it has been increasingly difficult to retain such a clear dividing line between state guidelines and local responsibility, particularly if central government participates in a policy project. This is very much the case with economic policies, because the various space economies are so closely interlinked. Therefore, in the interest of effective policy making, some form of cooperation and coordination between all government tiers is required. A simple, streamlined top-down approach is not available. Challenges to the role of local government, such as quangos competing with local authorities in planning matters, are uncommon because all political players accept the strong, constitutionally established local self-representation.

At the same time, the nature of local government finance encourages and, indeed, rewards competition between local areas both for businesses and residents, because they are both linked directly to the two main sources of local tax revenue: business tax and share of personal income tax. As a result, the more localist policies are, the more (financial) independence they may produce. This differs from the English situation with a stronger dependency on central government grants, little local financial autonomy and thus a weaker link between local policies and finance. This situation has not changed with the new regional structures which are primarily centrally funded. In Germany, the business tax is the most prominent and thus politicized case in point, because it provides a direct link between the performance of local businesses and locally generated public income. It is therefore politically tempting and rewarding to link policies to economic performance. The direct link, however, also harbours the considerable disadvantage of being pro-cyclical. Based on both capital values and the profits of local companies, this tax is very sensitive to changes in economic circumstances and each company's performance. A weaker economy thus provides fewer *local* means to take countermeasures. On the other hand, low tax rates may be set to attract new businesses, thus trading off improved prospects for economic development against financial scope to implement economic policies independently. This problem has stimulated discussions on alternatives, possessing the same degree of *localness*, but with, as yet, little outcome.

Financial provisions for local government in England and Germany show important qualitative differences, but also similarities. Both systems are based on central government allocations and local levies. Differences include the relative importance of locally managed sources which offer scope for autonomous local policy making which includes the regional level through interlocal networking and jointly funded initiatives. At the same time, however, the strong dependence of municipal budgets on local economic performance introduces a distinct pro-cyclical quality. In times of a weakening economy, which requires strong policy responses, dependence on central allocations increases, leaving local self-government in Germany less autonomous than statutory regulations might indicate. Although this suggests some similarities with the situation in England, the difference remains that *Kommunen* in Germany possess guaranteed sources of revenue to cover running costs, and they also have more discretion in budget management, than their English counterparts. In England, local budgets are subject to more detailed central control and frequent changes to operating conditions and regulations, depending on the political objectives in Whitehall. Thus, English local government faces detailed and unrestricted central control and, if seen appropriate, direct intervention, whereas in Germany a sphere of local discretion is guaranteed. This suggests a different degree of local confidence, but also concern about any threats to that independence.

Federalism, regionalization and institutional competitiveness

Germany's federal structure provides for a strong bias towards the meso level of government, because of its devolved, federal structure. Power rest with the *Länder* as mediators between local and national government tiers (Wollmann and Lund, 1998). Each *Land* has the possibility and, as part of the planning hierarchy, statutory obligation, to establish regions as subdivisions of the *Länder*. They include administrative regions, such as the territories of the *Land* Government Offices (*Regierungsbezirke*), or the planning regions, subdividing the *Länder* into several planning areas. Effectively, therefore, two types of regions exist with considerable institutional and scale differences: the *Länder* as centres of regional government and the Planning Regions (*Planungsregionen*) as mere territorial containers to administer planning policy. There has thus been, traditionally, a distinct top-down approach to regionalization with the main focus on hierarchically organized spatial planning. The federal government for the most part takes a back seat, providing strategic guidance in national development aims and collaborating with the *Länder* in the joint financing of regional development policies, including those funded by the EU (see also BBR, 2000 on Germany's planning system). This situation is very diffe‚rent from Britain where all such policies are administered by the national government, albeit with similar policy objectives.

This centralized perspective is juxtaposed to a strong, self-conscious local government and its in-built interest in regional matters both through the close functional interdependencies between local and regional matters, e.g. public transport networks, and the established two-way system of a central–local interrelationship which encourages bottom-up relaying of local concerns. Effectively, therefore, regional considerations are developed in a power field between a strong local government and the *Länder* as central government with their keen interest in using 'regions' as a device to implement their own policy goals at a smaller scale and to exercise a degree of control on the contents of local planning and development policies. The ambivalent nature of regions between central and local interests, their vague statutory description and *Länder* acting as central government, make regionalization open to *Land*-specific interpretations of what constitutes a region, how it is delimited and how operated (bottom-up or top-down). Not surprisingly, the result has been a rather 'heterogeneous regional level' (Benz, 1998: 128) with, inevitably, somewhat unclear notions of what constitutes a 'region'. Since the 1964 Planning Act (*Raumordnungsgesetz*, ROG) a top-down approach has been firmly established, requiring the *Länder* to create formal planning regions as further subdivisions of their territories. These constitute the statutory regions which serve as the basis of regional planning and policy within the institutionalized planning hierarchy. These planning regions, about three to five in each *Land*, develop regional plans through their Regional Planning Associations (RPA) which consist of members of the participating local authorities. Depending on the degree to which a *Land* allows regional policy to be a matter of interlocal cooperation rather than top-down initiative, the planning regions are either attached to the regional *Land* representations (Regional Offices) or held at arm's-length to allow greater local input.

The RPAs have no constitutional powers and operate through the participating local authorities, which also provide an indirect democratic legitimacy, and they offer a good bargaining base towards the *Land*. All local authorities are represented either directly or through groupings of small authorities (*Kreise*) (von der Heide, 1994). The RPAs are, by their nature, quite comparable to the new regional chambers in England. Inevitably, given the unevenness among local authorities in size, institutional capacity and available leadership capabilities, the urban authorities, especially the metropolitan regions, have a strong influence. Frequently, this urban–rural contrast leads to political controversies between the specific interests of the two sides.

This point has repeatedly been made during interviews with local and regional policy makers in the years since 1997.

In Germany, not surprisingly, the economically stronger localities seem to be more influential in shaping their regions, whereas the rural authorities seem to be less sure about regional interests and identities and to what extent they represent a possible infringement on their autonomy. The 1998 review of the 1964 Planning Act indicates a shift in emphasis towards more informal, network-based (mainly involving cities) and locally defined regionalization. This is the institutional acceptance at national level of policies and projects developed by *Land* governments as an alternative to the traditional hierarchical model. Such alternatives include the less formalized Regional Conferences pioneered in North-Rhine Westphalia as platforms of informal regional collaboration between (largely industrial) agglomerations (Heinze and Voelzkow, 1997), or the *Regionalverbände* (regional associations) in Baden-Württemberg and Hesse, as exemplified by the arrangement for the Frankfurt region (see Chapter 4).

Such informal approaches to regionalization are now also gaining attention in eastern Germany, for example in Saxony-Anhalt, with the formation of Regional Conferences and Regional Forums as informal, network-based collaborations between groups of local authorities which have identified common interests. In fact, experiences with the initially top-down regionalization across eastern Germany have encouraged the review of regional planning policy. In Saxony-Anhalt, as discussed in Chapter six, four new, locally defined and network-based 'regions' are now being institutionalized as official planning regions. They replaced the three regions which were established initially by the *Land* simply by declaring the territories of the three *Land* government offices as planning regions. The urgency with which new institutions were required after unification made such a heavy top-down approach seem appropriate. As one might expect, the formation of these formal planning regions could therefore not always be very responsive to existing cultural, historic or geographic identities, less so than was the case when the new *Länder* themselves were established. This may suggest that, at least initially, planning regions were considered more of an administrative rather than governmental feature. Also, there was little sensitivity to economic territories, as the administrative divisions across the economic region of Leipzig and Halle demonstrate. Here, the two *Länder* Saxony and Saxony-Anhalt had to agree a state treaty to enable cross-border cooperation in regional planning to catch up with federal links.

The changing circumstances of economic regions and the need for a more flexible approach to policy making at that level, e.g. through city networks (Gleisenstein, Klug and Neumann, 1997), have been recognized by the federal government. It is interesting that such a process of envisaged 'deformalization' is driven from the top with the aim to establish 'flexibility and informality' as an integral part of institutionalized, formal, regional policy. A particular feature of these city networks is their inherent dynamism, allowing new alignments to be established quickly in response to perceived (temporary) common interests and policy objectives. These include, in particular, the development and portrayal of regional images as part of more aggressive area marketing, but also the development of joint initiatives, such as transport (Göppel, 1993). Such a new approach to regionalization has been adopted by the *Länder*, for example in the northern western *Land* of Lower Saxony where the new structure is part of its economic policy to compete more effectively in international markets (Krumbein, 1997). This new approach includes a variety of interlocal collaboration with different degrees of formality: regional associations as non-profit organizations, which include participating local authorities and other, non-government actors. Then there are *Strukturkonferenzen* as very loose agreements between participating local authorities to meet to discuss development policy. There are also 'regional forums' as working parties consisting of non-urban counties (*Kreise*) and also attempts at collaboration across *Land* borders, such as between Saxony-Anhalt and Thuringia.

These various forms or organizing collaboration show the somewhat experimental stage of this new approach to regionalization. It seems the main emphasis is on setting up these new arrangements, while policies and strategies seem less obvious (Krumbein, 1997). A similar *Land*-inspired shift in regional policy has been adopted by Thuringia, in the Thüregio Model, which was developed by the *Land* government and thus clearly illustrates a top-down approach in regionalization, in policy, structure and operation. The main focus is on formulating regional development concepts as a joint interlocal effort to develop a comprehensive, integrated, planning and marketing oriented development concept (Hosse and Schübel, 1996). The range of different forms of regional networks suggests different expectations of such arrangements in terms of their aims, strength and transfer of local powers and responsibilities. The informal nature of these constellations means that no clear framework exists for their operation that has been agreed by local and *Land* government, both of which claim responsibility for regional matters. This is the case with the Regional Conferences in North-Rhine Westphalia, where the *Land* views them as forums of regional cooperation, while local authorities see them as a base for bundling their political bargaining with the government (Fürst, 1994). The shift towards a new form of regionalization has also been facilitated by the federal government as part of a national competition 'Regions of the Future – Regional Agendas for Sustainable Spatial Development' (BBR, 1999a). The result has been a diverse range of initiatives, with differing emphasis on project numbers, coherent development concepts and critical appraisal of strengths and weaknesses. One of the main outcomes has been the raising of public awareness and discussion between actors in the regions and this stimulation of a regional awareness. How far this goes to facilitate genuine collaboration and effective policy making at the regional level remains to be seen. What has become certain, however, is the realization among policy makers

that regions and regional policy matter increasingly for international competitiveness both at the local and national level. This includes the recognition of European Metropolitan Regions which, since 1997, have become an
official part in Germany's national planning paradigm (Voelzkow, 2000).

Regionalization and spatial planning

There are indications in both countries that regionalization can operate
through different avenues of region-building: territorially fixed and institutionalized from above and informally with flexible boundaries from below.
This reflects the position of the region between the local and the national
level both in geographical and institutional terms. The lack in clarity of the
nature of regions, their role and purpose within state governance, has provided for a range of institutional models of regionalization. Constitutional
and statutory regulations are obviously crucial in providing the framework
for the respective institutional machinery and provide territorial fixity through
institutionalization.

Existing administrative boundaries and, particularly in eastern Germany,
political sensitivities about references to old East German territorialization,
have influenced the drawing of regional boundaries by the *Länder*. The
picture of regionalization is thus varied, circumscribed by a *Land* government's willingness to allow locally-led regionalization to become 'official
policy'. Nevertheless, there seems to be generally a growing presence of
more informal, network-based forms of regionalization, irrespective of the
forms of official *Land*-defined regionalization.

The trend towards a more open, flexible understanding of 'region' is
reflected in a growing number of Regional Development Concepts or
Regional Conferences, as in the *Länder* of Saxony-Anhalt and Saxony. This
move was encouraged nationally by the federal government. These are strategies to facilitate interlocal collaboration within functionally defined regions
Such concepts are not part of the formal planning and policy-making instru-
mentation, but represent a new initiative to facilitate regional economic
audits as a basis of regionally defined, and explicitly economy-related, policies
(Scholich, 1995) to utilize indigenous potential. The focus is on collaboration between local authorities to develop regions that are more homo-
genous in their social and economic geography, and possess greater functional coherence, than could be expected from a centrally-defined 'artificial'
region (Schmitz, 1995). Each *Land* has followed this road to new regionalization in its own way. Differences largely concern arrangements for the
dual nature of the official planning regions between *Land* control, exercised
through their regional offices, and locally based planning associations,
consisting of delegates from the unitary urban authorities and higher tier
local authorities (*Kreise*). The latter form of regionalization promises a
stronger regional identity than could be expected from a conventional

top-down defined regionalization (Schmitz, 1995), often transposed from western Germany to the eastern *Länder*, rather than developed *in situ*.

Traditionally, not least for historic reasons, regions and regionalization have played an important part in Germany's system of governance, institutionalized in the form of the *Länder* (federal states) and, at a smaller scale, as planning regions outside the hierarchical system of spatial planning. While the former exercise considerable governmental power, the latter are mere administrative entities, defined by the *Land* governments. They are now subject to discussions on regionalization between local and *Land* levels, much facilitated by developments in post-unification eastern Germany over the last decade. These include emerging trends towards a more flexible, less territorially fixed and locally led approach to regionalization with a greater role for strategies rather than planning (Danielzyk, 1995).

Formal regionalization has been established as an integral part of the spatial planning system by federal law (1964 Planning and Development Act, *Bundesraumordnungsgesetz*, BROG), and includes three main tiers which correspond to the governmental hierarchy in federal Germany:

1   the national (federal) level with its nation-wide responsibility for strategic development planning as laid down in the Federal Spatial Development Programme (*Bundesraumordnungsprogramm*).

2   the *Länder* (federal state) as the main regional level of government are responsible for two types of 'regions' and their planning:

    i   development plans for complete *Land* territories (*Landesplan*) as strategic frameworks for local development planning, and

    ii Regional Plans (*Regionalpläne*) for the Planning Regions within each *Länder* as defined according to *Land* policy objectives and planning ideals. Planning regions are not part of the official hierarchy of territorial governance and serve solely administrative (planning) purposes.

3   the third level is that of local government with its exclusive responsibility for local development planning and control (see e.g. Kistenmacher *et al.*, 1994: 44ff) as part of constitutionally guaranteed local self-government (cf. e.g. Petzold, 1994).

Cooperation between these rather autonomous local policy-making entities at the regional scale is the main focus of discussions on more flexible, locally based and essentially modular regionalization. On this basis, planning regions would no longer be territorially fixed as a basis of medium- to long-term spatial planning, but would exist as (temporary) groupings of local governments with common regional interests. The result may be 'Regional Development Strategies' (Danielzyk, 1995) or Regional Development Concepts as outlined in the last review of the BROG in 1997.

The unclear statutory provisions made for formalized regional governance have resulted in various interpretations of the respective roles of *Land* and local governments and their underlying aims (Schmitz, 1995) between *Land*-directed, top-down forms of regionalization and a more locally driven, bottom-up style of regional initiatives. The former operates 'regions' in the sense of regionalized *Land* development plans and strategies, while the latter views regions as the result of voluntary collaboration between local governments. The effectiveness of such 'bottom-up' pressure will depend on local institutional capacities, and the abilities to utilize existing powers. Both were rare among the newly empowered eastern German local authorities used to an autocratic socialist system (see also Biskup, 1994). This encouraged or even required, at least initially, a stronger *Land*-led approach towards establishing 'regions' as part of the West German-style government structure. The result has been different approaches to regionalization, pointing to the embarkation on *Land*-specific 'pathways' of development (Müller, 1995).

There are four levels of spatial planning, corresponding to the three main scales of government in Germany. Listed with their main characteristics, the four tiers of planning are shown in Table 5.1 (after BBR, 2000).

Local planning control is one of the main local powers and responsibilities as part of local autonomy. The execution of these powers fundamentally depends on local finances, especially those generated and controlled locally. With local taxes among the main sources, next to central government grants, and both dependent on population number, the size of communes is crucial for local policy-making capacity. This will also decide on their position (role) within regional policies, e.g. as part of city networks. It is here where great differences exist between eastern and western Germany, with the former showing a comparatively small-scale structure and thus reduced local capacity both institutionally and financially (Figure 5.4). The average per capita local revenue of the 'top ten cities' which are all in the west ranges between 2,100 and 3,700 DM, while the bottom ten cities largely in the east, generated a mere DM 450 to 560 in 1999 (BBR, 2000). The much smaller scale of eastern local authorities compared with their western counterparts is one of the main reasons: three quarters of all municipalities in the west have fewer than 5,000 residents, while this share is in excess of 90 per cent in the New *Länder* in the east (Table 5.2).

Urban planning operates through a combination of formal and informal instruments. The main formal instruments are: land-use planning (general structure plans and detailed local plans establishing local building right), urban development and renewal plans (selected zone/areas designated for special policies).

The intense competition for funds and investment has facilitated a debate on regions and regionalization. This is important, because of the very different qualities of understandings of 'regions' and their nature, and the meaning of 'sustainable'. The regional planning association of North

Table 5.1 Planning responsibilities at government levels in Germany

Government level	Federal	Land ('Macro-regions')	Region (sub-Land level)	Local (municipalities)	Public and private organizations
Characteristics of plans and planning	• comprehensive competencies • general and vague phrases, little detail • basic ideas and strategic goals • issues guidelines for Land regional planning	• specify Land-specific development instruments and objectives (e.g. development nodes, axes, etc.) • establish regional planning legislation, programmes, plans • defines building regulation	• operate planning regions • detailing Land planning goals through regional plans for the planning regions • no corresponding tier in the government hierarchy exists!	• autonomous control of land-use planning and building control • two-plan system: local plan and structure plan (with growing detail) • establishing local planning law	• planning and development of individual projects upon 'invitation' (contracting in) by local government

Table 5.2 Number and population size of municipalities in eastern and western
Germany

Inhabitants	< 5,000	5,000–20,000	20,000–100,000	>100,000	All munici-palities
Eastern Germany	5,275	322	95	12	5,686
Western Germany	6,143	1,785	494	70	8,510
All	11,418	2,107	589	82	14,196

Data source: BBR (2000).

Thuringia views for instance, any initiative as 'sustainable' which benefits
the region, because it affects the region's economic viability. 'Sustainability'
can also serve a unifying purpose, bringing together diverse interests in a
region. This is the case, in particular, when there is no tradition and together-
ness felt in the region, such as North Thuringia. Competition can help focus
the minds (BBR, 1999a: 42–43).

Regions and regionalization in Germany in the 1990s: from regional plan to city networks

The 1990s saw considerable shifts in the public understanding of regions,
regional development and policy, facilitated *inter alia* by the challenges of
German unification in 1990. Rapidly growing and changing disparities at
regional and subregional levels highlighted the need for a more flexible
approach, both territorially and institutionally, than the traditional instrument of formalized planning could offer. As a result, government has
adopted the notion of networks as a basis of strategic reasoning and practical planning implementation, which is presumed to be in a better position
to develop and deliver solutions to regional development problems. Its main
advantage is seen in identifying and bundling common interests between
actors at different scales of government (BMRBS, 1996: 9). The decentralized organization of the settlement structure in Germany, and its multilevel
government system, support further decentralization of a traditionally more
hierarchically organized planning political system. Effectively, the new
emphasis reinforces the established 'principle of counterflow' (see above)
between the various government levels as part of a policy definition process.
The new regionalism shifts the emphasis more towards a bottom-up process, while also enhancing the role of the regional scale as 'intermediate'
negotiation and bargaining sphere between federal and *Land* government,
and local government. The federal government is instrumental in pushing
regions into that new role with its revised planning and spatial development

legislation, which emerged from the Ministerial Conference for Spatial Planning in 1997 and the review of the Federal Planning Act.

> Thus, regional planning does not end with making plans and regulating development, but evolves into a dynamic political process of mediation on the goals for regional development and their implementation. This exceeds the traditional concern with creating perfect systems of regional plans, by far. In future, the main emphasis will thus be on finding common solutions between central and local government and also private actors. Planning no longer is to be conceived as state control [Staatsaufgabe] but the provision of a service, able to cooperate and coordinate.
>
> (BMRBS, 1996: 10, author's translation)

This shift, however, may just as well raise questions about the very nature of regions, their territoriality and scale, as these will fundamentally influence the number of regional actors and this, in turn, the range (and potential diversity) of issues. Territoriality may thus translate directly into negotiability and the capability of reaching a compromise. Implicitly, this issue is being addressed by the emphasis on flexibility which will gain in importance with the number of players. The new emphasis on networks and cities as the cornerstones of regions as potentially flexible territorial constructs sits somewhat uneasily next to the continued engagement with regional plans which, by their nature, are static and inflexible. It thus seems like attempting to straddle two stools: continuing traditional formal planning while facilitating flexible collective bargaining and interlocal alignments.

The main goals of traditional spatial planning at the federal level now include incorporating the new understanding of a more flexible, less formalized, regional planning. This means a continued, and renewed, emphasis on developing a decentralized settlement structure around city networks to complement the existing, traditional central place-based system. 'Decentralized Concentration' is the new buzzword, together with 'cluster' and 'network' (Elsner, 2000) The envisaged urban networks require continued improvement and modification of relationships and linkages. Their role and operation and scope have been studied in a federal research programme 'City Networks' as part of the Experimental Housing and Urban Development scheme.

Germany's federal structure requires the interoperation of various government tiers, while maintaining a hierarchical structure from federal to local government. The elements of spatial planning are (BMRBS, 1996):

1   The Federal Planning Act ROG of 1965 which establishes the statutory framework, including the distribution of responsibilities between government tiers.

2   Tiers of planning corresponding to the state hierarchy, and their inter-relationship as key mechanisms of the system.

3   Meeting of *Land* Ministers for Planning and Development as a bargaining and coordination platform (*Ministerkonferenz*) vis-à-vis the federal government.

4   Programmes and plans. While the former are essentially strategic papers on development aims and objectives, the latter comprise technical and legally binding plans. There is one development programme for each *Land* (*Landesentwicklungsprogramm*) containing aims and objectives for the *Land* territories. The plans also establish smaller planning regions as spatial containers of more detailed (*Land* plan-guided) regional plans as a product of local and *Land* government cooperation.

The revised version of the Development Policy Framework (*Raumord-nungspolitischer Handlungsrahmen*) of 1995 provides specifically for the creation and operation of city networks. It reflects experiences in post-unification Germany and the outcome of the 1992 Rio Summit on Environ-ment and Development and the subsequent promotion of Agenda 21 as the main paradigm for local and regional development. These have gained partic-ular attention in eastern Germany and, especially, in those cases where they link cities on both sides of the former Iron Curtain, such as between Saxony and Bavaria. The Saxon–Bavarian City Network is part of the model project City Networks supported by the federal government (see BMRBS, 1996). Infrastructure support to these networks is one priority of federal govern-ment investment.

The Framework for Spatial Development Initiatives gives more explicit guidelines to the *Länder* about required/envisaged initiatives in spatial development. In its 1995 version, there is clear support for the regional scale as a policy and planning arena, together with an emphasis on moving beyond merely making plans (section 1), and instead engaging with development processes and the mediation of conflicts. Recommended instruments include Regional Conferences as platforms for politicians and other actors to meet and discuss regional development problems and initiatives. Regional Conferences are generally little-regulated informal platforms for local and regional policy makers to discuss and find compromises and outlines for common regional strategies. There are no enforcement powers. Policy imple-mentation relies on the goodwill of regional actors both within and without government. Developed paradigms and objectives feed into a Regional Development Concept which is binding for regional actors. They thus create their own, eventually binding, frameworks, and so will be unlikely to include too many detailed provisions and, therefore, potential restrictions to local policies. Regional Conferences go back to the late 1980s when this approach was pioneered as part of a move to regionalize structural (economic) policy. Fifteen regions were established, roughly in line with the regions used by the chambers of commerce. The *Land*, however, retained financial control of grants and other investment decisions. Implementation follows either a joint regional approach through a regional development office, or it uses existing administrative structures, e.g. local government. Currently, all *Länder* have adopted this approach (Zarth, 1997). This fact reflects the realization that the existing formal territorialization of government does not necessarily match the developed pattern of functional territories and related policy requirements. Common interest and purpose and/or existing identities and affinities are recognized as crucial for establishing collaborative activities towards a common regional approach. Criticisms of this informal approach include claims of possible duplication of efforts, because formal regional planning would require such informal negotiating practice anyway (Schädlich, 1997).

Regional Development Concepts are nevertheless seen as useful instruments to formulate 'joined-up' policy goals. City networks are recommended for their strength as flexible and problem-solving devices. They are not seen as a new tier in the established planning hierarchy, but rather as a form of voluntary cooperation complementing the Central Place System as the official, institutionalized and centrally directed functional relationship between cities (Göppel, 1993). However, there are signs that this 1970s' paradigm is being reassessed for its continued appropriateness (Elsner, 2000). City networks and, since 1997, metropolitan regions are the new emerging concepts in regional planning, injecting flexibility and dynamism into the static and potentially cumbersome formal system of regional development planning and Christallerian relationships.

The new approach has been 'test-run' in 11 city networks across Germany as part of a federal competition for best practice, and seeks to capture the various regional issues, such as east–west contrast, cross-border, urban–rural and metropolitan (industrial) region, Iron Curtain, such as between Saxony and Bavaria. The Saxony–Bavarian City Network is part of the model project City Networks supported by the federal government (see BMRBS, 1996). The main objective of this project was to identify the organizational form of spatial development planning most suitable for responding effectively to rapidly changing circumstances, while improving regional competitiveness (BBR, 1999a). Despite their differences, there are common themes to the policy goals of the city networks: improved accessibility (infrastructure), city-regional marketing and image making, and economic development and competitiveness. Among the 12 city regions participating in the 'best practice' programme, tourism was mentioned ten times, transport/communication 11 times, technology centres and transfer six times, and city-regional marketing five times (BBR, 1999a: 96). Evidence so far is that city networks have contributed to facilitating interlocal cooperation. Furthermore, they provide further instruments for regional planning and policy making, which enhance local capacity (through the sharing of responsibilities and expenditure) to

Table 5.3 Adoption of city networks as instruments of regional planning and policy in the German *Länder* post-1998

Land	Position of city networks
Baden-Württemberg	Adopted in 1999/2000 as official policy with emphasis on local–regional cooperation.
Bavaria	City networks as contribution to improve international competitiveness, supporting established land development programme.
Brandenburg	Cooperation between Berlin and surrounding Brandenburg as 'joint task' in regional planning for immediate hinterland of Berlin.
Hesse	City networks as complementary to existing policies with emphasis on resource bundling and cost savings.
Lower Saxony	Vague description of city networks as 'regional collaboration'.
North-Rhine Westphalia	Broad acceptance and adoption of city networks as part of regional planning. *Land* policy is to be aimed at facilitating and supporting networks.
Rhineland-Palatinate	City networks as strategy to enhance competitiveness. Emphasis on problem-oriented regional policy in response to city networks, reference to central place structure.
Saxony	Formal and informal city networks introduced. City networks are promoted as special case of general interlocal cooperation, which generally is aimed at increased efficiency in land use and financial resources.
Schleswig-Holstein	City networks as part of *Land* regional strategy, with main emphasis on mobilizing endogenous economic potential.
Thuringia	In process of adoption into *Land* Development Strategy.

Source: After BBR (1997).

pursue policies and use resources more economically. They also counteract narrowly focused localism. The voluntary and egalitarian participation of localities is essential in this initiative. It also has emerged that city networks help to overcome borders and facilitate the inclusion of non-governmental actors as part of a shift from government to governance.

Formal regions and flexible regions in England and Germany – some observations

On the face of it, the English and German experiences present very different contexts for regional governance. However, in both countries change is a constant theme. In both there is a new concern to get the regional scale right. In recent years we have seen the emergence of more flexible, locally based regionalization in Germany, although these processes sit alongside top-down regional policy and planning. The centralizing tendency of English government was, in the 1990s, modified by first deconcentration of government offices and second by the formation of Regional Assemblies, Chambers and the RDAs. However, the extent to which this represents an acknowledgement of bottom-up pressure or part of the central state's approach to increased economic competitiveness raises unresolved questions. The evident frustration on the part of government that the RDA reform had failed to overcome coordination problems brought the Cabinet Office at the heart of government into the management of regional policy. In Chapters 6 and 7 we continue to pursue the tensions in regionalization in both countries by examining a series of local experiences of these national trends.

6 Monocentric city regions in unitary and federal states

Experiences of regionalization in England and Germany

This chapter and the following present four examples of regionalization processes, comparing monocentric and polycentric regions in the very different institutional, practical and historic contexts of the government systems of England and Germany. This includes in particular the varying territorial scales at which regions have been pitched between the local and national. The cases illustrate the obvious differences in forming and operationalizing regions, against the shared sense of needing to respond with appropriate regulative arrangements to the perceived growing regionalization of the economy. The nature of these responses varies considerably, shaped by area-specific characteristics, national government structures and established provisions for city-regional governance. We examine the importance of such variations and determining factors in relation to the differences in the wider picture. This wider context includes national constitutional particularities, attitudes to devolution and subnational policy-making autonomy within the national state frameworks and the responses to the changing economic reterritorialization of the space economy in favour of the regional scale.

This and the following chapter thus explore the combination of cross-national and cross city regional comparisons (Figure 6.1), exploring the respective importance and operation of differently scaled institutional structures. In a period when regional institutions are conscious of the need for improved competitiveness in a globalizing economic environment, interesting cross-national and cross-regional similarities emerge that transcend formal and historical differences. In all of the cases informal linkages between governmental actors and between governmental and non-governmental actors seem to be significant sources of difference. Comparison of the cases is not therefore based on simple national differences, but contrasts the more specific determinants of regionalization processes. Each of the city regions seeks to find its own way towards an appropriate institutional response, evolving between its general constitutional-institutional framework and its particular local–regional circumstances.

The four case studies illustrate four scenarios of varying combinations of national (external) and city-regional (internal) contexts (see also Chapter 4),

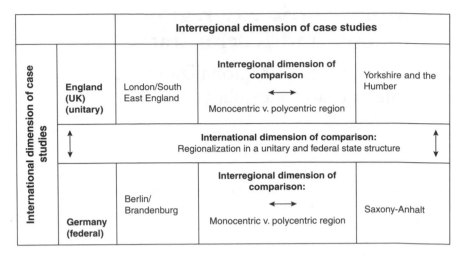

Figure 6.1 Inter- and intranational comparison of case studies.

each with their particular emphasis on the city or the region respectively (Figure 6.1). Thus, there is the national dimension with its fundamental differences in devolution of powers to the regional and local government levels, between the federal and unitary systems respectively. The other main difference rests in the balance between the local (urban) and regional interests and institutional presence. Thus, for both countries, a mono- and a polycentric region have been chosen, because of the expected impact of internal relationships which form the *city* region or the city *region*. It might be assumed that the polycentric city regions would reveal greater inherent regionalizing dynamism, based on the less dominant role of the urban centres compared with the monocentric city region, and a greater sense of gaining competitive advantages against other regions. The smaller cities, individually less dominant in economic terms, make regional collaboration seem an advantageous strategy that increases local visibility and influence in national and global economic competition. At the same time, however, they may feel compelled to engage in competition for immediate political or, as in Germany, fiscal gain, thus effectively ruling out collaboration within the region. The opposite situation could be assumed for the monocentric city regions, where the dominant presence of one urban economic centre may be expected to reduce the rest of the region to little more than its hinterland which may not necessarily be seen as an automatic gain by the city. The four case studies thus allow us to compare and contrast the relative impact of external, national and specific internal, city-regional issues.

Installing new regionalization in the London region

London and the south-east of England illustrate a monocentric region which has seen fundamental institutional changes as part of the UK government's devolution agenda. As a result, a number of new regional institutions increased the complexity of governance, with potential for competition, and the new regional territoriality resulted in new boundaries dissecting the south-east region and, crucially, separating London from its hinterland. The London region is characterized by uneven development with strong economic and social divisions (see Allen *et al.*, 1998). A number of areas within the London region have effectively been excluded from continued growth with its associated pressures on land and environmental resources. The capital region sits within the core of the European economy. London, however, has also quite distinctive characteristics as a world city that distinguish it from most other European cities (see Le Galès, 2000; Taylor and Hoyler, 2000; Graham and Hebbert, 1999). The importance of London and the south-east for the English and UK economy is beyond doubt, having underpinned the growth-redistributing regional policies periodically introduced by central government. Generally, national economic policies have tended to favour development in south-east England, and national fiscal policy seeks to manage periodic overheating of the regional economy. Development pressure concentrates on central London and in recent years along the western corridor beyond Heathrow to Reading, but there is also considerable interest in, and by, smaller towns in south-east England for further development for both housing and business. At the same time, however, there are pockets of deprivation and economic underachievement within the area. Government policy has historically sought to counteract the development pressures on the western part of the London region and redirect them eastwards, but such aims have only relatively recently begun to be realized with the success of redevelopment in London's Docklands. Current policies support growth further eastward along the Thames Gateway, along the London–Cambridge corridor and encourage planned expansion around Ashford in Kent and the Medway Towns (Map 6.1) (DoE, 1992, 1995). All of these initiatives reflect considerable direct involvement by central government in the development of the region. European funding has also tended to concentrate on the areas east of London, such as south-eastern Kent. Outside London, the economy reaches out into two newly formed English regions that cover the area previously called Rest of South East (R.O.S.E.) and parts of the former East Anglia region.

In addition to its complex economic structure, the London city region is governed by a complex set of institutions. Perhaps the most distinctive feature is the new office of a democratically elected mayor to represent Greater London. This position makes the new region of Greater London different from all other English regions. The London economy, of course, extends beyond the 1963 boundary of Greater London. The city-regional

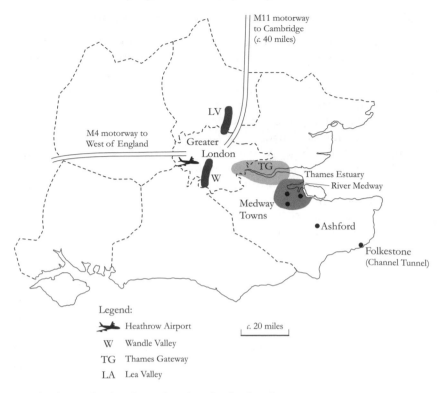

Map 6.1 Strategic planning subregions in the London area.

economy stretches over three formal regions. The new regions then include a large number of local authorities with overlapping economic development and planning responsibilities. In London, there are 33 local authorities plus the office of the mayor and the Greater London Authority. In the South East region alone there are 74 local authorities, including seven county councils and 12 unitary authorities.

Irrespective of all these changes central government continues to have an important stake in the governance of the London area. Government institutions include a Minister for London and three government regional offices responsible for the newly created tripartite territorialism covering the former South East region. In addition, there are a number of subregional divisions and also divisions of responsibilities between local governments and quangos. Eligible areas for EU funding impose further internal boundaries within the regions Lea Valley (see Map 6.1).

Much is new in the insitutional landscape. But change, especially in the case of reform in London, also reflects long-standing tensions in intergovernmental relationships. The history of London government is one of change initiated and managed by national governments (see Travers and

Jones, 1997). Reform in the 1960s, which created the Greater London Council, joined inner and outer boroughs under a single authority, better reflecting the spatial economy, but with the political objective of securing Conservative majorities for the new GLC. In the event, political power shifted between the main parties and the period of Labour control in the early 1980s provoked a Conservative central government into further reform and abolition of the Greater London scale.

The new government of London responds to strongly expressed popular demands for a London-wide democratic level, but in the context of central government's wider devolution agenda. As pointed out earlier, devolution to Scotland, Wales and, especially, the English regions has been limited, and in England central government retains a strong hand on local and regional government finance. Much of it stems from a deep-seated distrust of sub-national governments' ability to manage their affairs effectively. Thus, while elected government in London was widely supported there were early doubts about its effectiveness,

> such new institutions risk being badly disabled if they do not have suffi-cient political and financial power. There is a likelihood, in a country as centrally run as Britain, that no government would give London the effective government its electorate desires.
>
> (Travers and Jones, 1997: 39)

As we shall see, the limited devolution in London creates problems for effective management of the city region. What most studies of London government in the early 1990s identified, was demand for a democratic voice and also a need for strategic coordination (Newman and Thornley, 1997; Travers and Jones, 1997). Democracy proved to be a problem. On 4 May 2000, Londoners voted for both a mayor and the London Authority. This part of the Labour Party's devolution programme, however, did not go according to plan. Not only did Labour fail to get its candidate for mayor elected but it also failed to win a majority of seats in the GLA. In effect, no single party won a majority of seats, prompting Labour and Liberals to form an alliance to elect committee leaders and exclude the Conservatives from official positions. In the weeks immediately following the elections the new mayor attempted to find GLA members who would work in his exec-utive. The political controversy surrounding the reform of the government of London should come as no surprise given the history of local govern-ment in the capital, and the inherent political tensions with the centre.

Just as the objective of democracy proved difficult, the objective of strategic coordination in the capital seems elusive. The mayor and GLA add to London's governance institutions which include four new functional bodies as executive agencies for the GLA's main functions: the London Development Agency, Transport for London, the Police, and the London Fire and Emergency Planning Authority (Simmons, 2000). Both central government institutions – the Government Office for London and the Minister for London – continue, as do the lower level borough councils. Functions are distributed between these and a large number of other agencies. In the years after abolition of the GLC there was persistent criticism of the lack of coordination between the range of public, private and partnership bodies. As will be seen, this criticism has not been answered. This institutional problem was nowhere more evident than in relation to the transfer of responsibility for the capital's public transport. The controversy over the deterioration of the quality of public transport dominated the process of electing a new government for London. At the heart of this debate is the means of funding improvements to the Underground system. The Jubilee Line extension, opened in 1999, had not only incurred numerous embarrassing delays, but also run well over its budget. Initially, it had attracted some private investment from the developers of Canary Wharf, but this turned out to be a small contribution to the overall cost. The Labour Party was unwilling to commit public expenditure to renewing the Tube. The transfer of responsibility for the Underground was therefore delayed beyond July 2000 to ensure that whatever solution central government favoured could be put in place before the mayor took control. The mayor would thus be left with minor decisions and the day-to-day problems and complaints of Londoners about public transport, yet would have little freedom to respond to what was consistently defined as the biggest issue facing London. The transport issue reflects the continuing willingness of central government to devolve responsibility but not power to London.

Perhaps the greatest sense of continuity in London government lies in the continuing role of the Corporation of the City of London at the heart of the regional and national economy. Throughout the history of metropolitan government in London, the City has resisted reform. The City's medieval structure, electing a government from among business voters, resisted all attempts at democratic reform. Following the abolition of the GLC, the City took over many London-wide functions including responsibility for some parks, and assumed a role in lobbying on behalf of London. The current reforms also leave the City as a separate authority, but the mayor succeeded in bringing the leader of the City into his group of close advisors.

Intergovernmental relations in London are locked into historical conflicts between central government and the sub-central tier(s). However, new institutions have been added to the governance of London and the following sections will examine how new roles and responsibilities have begun to be exercised. Thus, despite the many flaws in the way in which the position of the mayor has been shaped, the mayor has substantial powers of appointment, strategic policy-making responsibilities and considerable power to steer debate. The effectiveness of these powers, however, will depend on the individual incumbent's personality and ability to use them strategically and discriminately. The mayoral position in London demonstrates the importance of personality in addition to the mere formal, institutional arrangements (see McNeill, 2001).

The London case will be developed under two headings. The first examines responsibilities for planning in London and across the wider region. Coordination between plans for London and planning in the wider region is a significant issue. The second heading includes relationships between the new institutions of economic governance in London and in the two new regions outside greater London. The dominant theme of this examination is the impact of administrative boundaries on effective policy making in one functional economic region.

Planning – technocratic cooperation in the London region

The image of a London region is fundamentally shaped by the strategic plan produced by Abercrombie in the 1940s, which addressed London and the south-east of England as an interdependent region of related development processes. While London was the growing core, the wider region was essentially London's hinterland serving London's needs. The Greater London Plan (1944) is the one which set the baseline for planners and established the idea of a London region in the regional planners' imagination. This marked out the ring of new towns and orbital transport, and set a regional context for London's development on the premise of managing and redistributing continued growth of London as the region's core, with the outer region offering complementary functions. Since 1962, SERPLAN (South East Region Planning Advisory Body) has advised and negotiated with central government on regional planning aims as *the* body seeking to bring together the region's differing interests between core and outer ring, and between east and west. SERPLAN was a voluntary association of planning authorities in the wider region including boroughs from within and without London. It offered a platform for informal, strategically oriented communication between local authorities. In effect, it represented some of the characteristics now being argued as typical of new, informal and issue-driven regional governance and informal region-based debate and cooperation. The continuity of city and regional cooperation was disturbed by the redrawing of regional boundaries in 1994.

Despite these visible territorial changes (see Map 5.1a,b), central government continues to have a strong hand in policy formulation expressed through Regional Planning Guidance to the two new regions surrounding London (DoE, 1994), and through approval of the London Spatial Development Strategy (SDS) for Greater London. Conceptually, the SDS is derived from the European Spatial Development Perspective (ESDP) and its spatially oriented strategic policy dimension. As such, the SDS provides a strategic tool to cut across the different functional roles of the GLA and the territories within London (Simmons, 2000). SERPLAN with the spatially much wider strategic perspective gave way to the Regional Planning Boards in the East and South East regions in March 2001. The increased formalization and functional divisions in strategic decision making across London's hinterland raise questions about the continuity of the established cooperative practices and expertise.

But such concerns should be seen in the context of a centralized style of regional planning through Regional Planning Guidance. Local planning authorities are consulted but the ultimate control and decision rests with the centre. Given the evidence of the diversity of issues and policy requirements inside and outside of regions, questions may arise about the likely scope for the centre to be able to respond appropriately. In the South East region the long-standing planning issue is urban containment and the tension between pressures for new development and protection of the countryside, often pitching London's interests against those of the wider region. In the mid-1990s, SERPLAN suggested a house building rate of 33,000 a year, but the latest RPG suggests 39,000 rising to 43,000 after 2006, with specific building targets imposed by central government on individual counties. While individual planning authorities may not want to accommodate this scale of development, their bargaining position is weakened by the abolition of SERPLAN. The government's proposals for reform of the planning system, published in 2001, suggested abolishing the county council role in plan making. Government negotiates with each planning authority, and individual authorities are reluctant to comment on issues about the distribution across the region and, instead, tend to stick to defending their own backyards. Effectively, this encourages localist perspectives and weakens scope for regional cooperation and the regions' standing in relation to local and national political forces and institutions. SERPLAN's functions passed to the three new regions, but in the new South East region, the assembly is unwilling to take over the debate on RPG9 which it sees as within SERPLAN's territorially wider remit and thus largely not relevant for its much smaller territory. This 'them' and 'us' mindset highlights the problem with policy continuity in ownership between territorially (and functionally) non-congruent institutions. Transitional arrangements thus present problems for planning coordination in the wider region and reflect the growing impact of subregional perspectives and interests. These coordination problems are recognized to a limited extent. The South East region, for instance, sees the main future challenges in finding ways in which the various bodies can work together 'in partnership inter-regionally to achieve common aims for London and the South East' (www.southeast-ra.gov.uk/southeast/ other.html, Oct 2001).

In London itself, new arrangements present another set of problems. Between 1986 and 2002, within Greater London, strategic policy was produced by central government taking advice from the London Planning Advisory Committee (LPAC) which was formally established in the mid-1980s after the demise of the GLC, and comprised representatives of the London boroughs. LPAC was a statutory joint committee set up to advise on matters of common interest for the planning and development of Greater London. LPAC produced two rounds of advice (the second compared favourably to Abercrombie's vision (Hall, 1994)) to influence the government's Regional Planning Guidance. One of LPAC's achievements was to encourage interborough cooperation on strategic planning in the vacuum left by the abolition of the GLC. No political party held a majority within LPAC and thus the organization had to search for consensus among the views of the boroughs. The LPAC period could be viewed as one of unprecedented cooperation.

Following election of the Mayor of London new formal arrangements for plan making and for taking decisions on large development projects have been put in place. The mayor has to produce a set of strategies, including those on transport and economic development, and also open space and waste management (cf. Simmons, 2000: 674–675). The Spatial Development Strategy (SDS) itself has a coordinating function (Simmons, 2000), but it is produced to a timetable allowing formal public consultation, and other strategies (the Economic Development Strategy was produced during 2001) can precede it and appear to compromise planning principles. In the early stages of strategic plan making the mayor had the benefit of substantial advice from LPAC. LPAC put together advice on a range of topics; housing, town centres, waste and minerals, the river Thames and the environment. Staff also transferred from LPAC to the GLA and thus helped transfer their expertise and working practices. The continued role of the former leader of LPAC as member of the London Assembly and other members of LPAC in drafting the SDS may help to maintain a sense of continuity. Thus, there was an opportunity for ideas developed in a period of close cooperation between the boroughs to be carried forward into the new strategic planning. The new mayor, however, was unwilling to accept advice not in tune with his election manifesto.

Perhaps not surprisingly, there was a tension between the mayor's political objectives and the formal plan-making process. Central government has given detailed guidance on the content of the SDS – it has to cover transport, economic development, regeneration, housing, retail development, leisure and culture, environment, built heritage, waste management, use of energy and London's world city role. In addition to this detailed central guidance, the LPAC legacy of interborough cooperation produces further constraint on the mayor. The mayor himself had picked up one or two significant planning issues – affordable housing and promotion of more office towers. Both relate to a policy continuity from the early 1990s that is the priority accorded to London's world city role. To 'test the waters', the early draft SDS, published in 2001 as 'Towards the London Plan' (GLA, 2001), set out political priorities, while avoiding the detail that would be required in the formal SDS (Thornley, 2001).

The SDS performs a statutory role in guiding the boroughs' plans in issues of common strategic importance. This formal relationship between mayor and boroughs replaces the cooperation enjoyed through LPAC. There are fixed procedures for consultation with the boroughs and central government, and for an Examination in Public. Strategic planning in these new arrangements sits between boroughs, the mayor, and central government, giving many potential opportunities for conflict. Conflicts in other policy areas – for example transport – are brought into the SDS process.

The issue that dominated the London policy agenda since the late 1980s was London's world city role. The London World City report commissioned by LPAC (Kennedy, 1991) set this policy direction, and GOL's commissioned Four World Cities report (Llewelyn-Davies, 1996) continued the theme. The competitiveness of the London economy was also prioritized by the London Pride Partnership (Newman, 1995). But will pursuit of this objective encourage a continuation of consensus politics and/or strategic planning in London and the wider region? Hebbert sees either 'global vision' or 'new localism' as possible futures (1998: 79) in response to the changed boundedness and set of tasks of the new institutions. This world city theme obviously needs careful balancing with social and environmental demands, but the remarkable consensus in both LPAC and the London Pride Partnership and its successors may not be sustained. The London Plan (GLA, 2001) starts by confirming the Mayor's world city ambition.

Even if policy priority to the economic development of a world city is agreed within London, there are significant issues of continuing cooperation in planning within the new interregional planning context in response to the newly divided territorial responsibilities in the south-east of England. The mayor is statutorily required to represent only London's views in the wider region, but to think about region-wide issues. Inevitably, this will add new separateness to, and between, the two regions surrounding London. For instance, there is no formalized platform for regular interregional meetings and policy discussions (and coordination). LPAC advised the mayor against formal interregional institutions and it seems that such meetings will only take place on an ad hoc basis for specific policy issues. This model is also favoured in other metropolitan regions to avoid commitment and possible constraints in local policy making (see Chapter 4). It is interesting to note that none of the three regions' websites make explicit reference to the wider region of south-east England. Almost to the contrary, the web pages on the South East Regional Assembly (www.southeast-ra.gov.uk) ask boldly 'What is the South East?', only to draw on basic statistical information on the formal South East Region. There is no mention of potentially different meanings of 'South East', as implied by the question. This suggests that wider regional interests and policy implications are not considered very relevant. The result is most likely to lead to spatially more fragmented strategic policy making and thus a greater need for coordination. This, however, will need to be outside the formal arrangements, where there is little provision for such connectivity. Interpersonal communication networks thus seem to be the only channels to cross the new borders in the wider

London region. In the ensuing competition of interests, London as the core city can be expected to push its case in relation to adjoining regions, a move that will most likely be resisted by the other two regions who want to be more than merely London's backyard and to maintain their separate development agendas. Thus, there is little mention of wider regional issues in the mayor's draft London Plan, nor is there any explicit, and certainly not prominent, reference to these on the GLA's web pages. In the months before the establishment of the new institutions in London, SERPLAN considered creation of a super regional organization including all planning authorities in the South East and East regions. LPAC advised against anything but minimal cooperation (LPAC, 2000). It did, however, recognize important cross border issues, in particular, for instance, for economic development along those economic corridors which cross formal regional boundaries. Coordination of economic development is, therefore, an important issue and as contentious as planning cooperation.

Economic development in the London region: basis of new competitive regions

Economic development and global competitiveness have been the main rationales behind the regionalization policies and act as the linchpins of the new institutional arrangements and the arrangements for informal cooperation between newly established institutions within the new regions. In all English regions, regional assemblies, regional development agencies and government offices, and in London, additionally, the high-profile office of mayor, add to a crowded field of economic development bodies. They are complemented by a number of non-governmental interest groups and organizations, such as the South East Cultural Consortium set up as a new body to coordinate cultural institutions and activities in the south-eastern region, and advise the Assembly and central government on region-specific issues.

Following the government announcement of its intention to set up the London Development Agency as one arm of the mayor's responsibility, a number of interests joined together in 1998 to set up an organization called the London Development Partnership (LDP). This organization was led by business interests and its purpose was to develop a strategy (LDP, 2000) that would then influence the LDA once it was established. The LDP grew from the business and public–private networks that emerged in London in the early 1990s (see Newman and Thornley, 1997). London First, including companies such as British Airways and the NatWest bank conducted research and produced policy papers, in particular about London's transport problems. Territorially, the wider London region was hardly visible. A link between business and local government was forged through the London Pride Partnership. The business leaders and the City Corporation contributed funds to the LDP to enable a small staff to be set up and various studies undertaken. The year's delay in setting up a regional development agency in London was thus offset by the public–private LDP. The new mayor, however, made new appointments to the London Development Agency and set new policy priorities. The chief executive of the LDA was one of the old GLC colleagues of the mayor. Business and the City continued to have access to economic policy, and consensus about London's world city role was expressed in the draft economic strategy. The close relations between the mayor and the business sector were also evident in the production of the draft SDS.

The dominant theme of economic development is competitiveness. RDAs in the neighbouring regions aim 'to take the strategic lead in promoting sustainable economic development' (www.seeda.co.uk), and use place promotion and regeneration as instruments of improving the regions' competitiveness (SEEDA, 2000). Some of these responsibilities have been scaled down from central government to the new agencies. The South East England Development Agency (SEEDA) states its mission as being 'to work with our partners to make the South East of England a world class region, achieving sustainable development and enhanced quality of life as measured by: Economic Prosperity, Environmental Quality, Social Inclusion' (www.seeda.co.uk). Similar strategy statements were made by the London and Eastern development agencies with 'three strategic "investment" themes: in business competitiveness and growth; in Londoners and their skills; and in communities and places' for London (www.go-london.gov.uk/ar2000/r.htm). The policy objectives in the eastern region are comparable. The aims and indicators used reflect the Government's influence on regional planning policies, primarily through the regional Government Offices.

The new arrangements provide for a division of responsibility horizontally, between the various regional bodies, including the development agencies and, vertically, through varying scales of government and regulation. Horizontally, there are Government Regional Offices, Assembly and RDAs vying with each other for influence on their respective regions' development. Inevitably, emphasis on policies, and institutional and financial capacity for effective involvement vary both horizontally within the regions, and vertically, between the tiers of government: *local* authorities with their continued broad range of functions, especially planning and development control and promotion (even if within central government parameters), *central* government with its budgetary controls and control of the European budgets.

The issue of global competitiveness is a recurring theme in all regional development strategies and also illustrates the multiplicity of actors engaged in relevant policies as well as the new divisions of territorial interests (and perspectives) in the wider London region. SEEDA's strategy identifies world class ambitions acknowledging that other European regions are outperforming the south-east (which SEEDA locates in 23rd position in a league table of European regions). SEEDA points out that a successful southeast benefits the country as a whole as growth centre and bridgehead into the economic sphere of mainland Europe. The economic strength of, and functional territorial economic interrelationships across the region overall, however, are not reflected in its new institutions. The spatially and politically diverse region has no obvious centre or alliance of political or business interests. Kent, for example, has stronger economic ties with northern France than with Oxfordshire. Indeed, Kent holds a separate subregional identity as a policy area within the south-eastern region. In addition to the territorially-based divisions, the Government Offices are also engaging in raising regional competitiveness as one of their key strategic aims (GO-East, 1999/2000 Annual report). Exceptionally, there is recognition of a wider territoriality. For example, in the RPG for the south-east region, an understanding exists of 'recognition of the region as essentially (but not exclusively) comprising London and its hinterland' (DETR, 2000), albeit limited to housing development. Such a broader view has not yet been reflected in the Region's own policy documents. It remains to be seen to what extent this diffuse geography and weak 'institutional thickness' in the south-east may be overcome by the process of devolution itself as the new regions seek to compete with the other English regions for private sector investment, and government and EU funds. New institutional capacities may be engendered by the enforced competition for funds on the existing arbitrary regional boundaries.

In the interest of successful economic governance, neighbouring RDAs need to work together. However, the rationale behind the RDAs was to make English regions more economically competitive, thus contradicting the notion of cooperation. In the competition for investment, London has identifiable leadership as the economic core of the region. The position for the south-eastern region is less straightforward in the absence of a clear political constituency or spatial coherence. Cooperation may be more likely between Regional Planning Boards of planning authorities, i.e. the formalized government-based regionalization, rather than the economic development oriented, business-led RDAs with their innate competitiveness. Wider cooperation may thus not be their first priority, but rather performing to government set performance targets (interview, SEEDA, Dec 1999).

In London, the absence of an institutionalized strategic body since 1986 has resulted in a number of alliances among interests based on territorial sectors of the London economy and established around common agendas rather than institutionalized territorial responsibilities. A set of subregional economies was identified in policy development in the 1990s. The 'wedges' – Heathrow/West London, Wandle Valley/Gatwick, Lee Valley and Thames Gateway (Map 6.1) – have complex histories of formal and informal cooperation originating in the early work of Business in the Community. Business Leadership teams, West London Leadership and the East London Partnership, pre-date the London-wide London First. It remains to be seen to what extent these networks can be brought into the new arrangements and how far the government engages in their facilitation and, indeed, survival of the new territorial boundaries between institutional responsibilities. In the wider region a differential economic geography can also be identified (Llewelyn-Davies, 1996). There are four groupings among the nine counties, with a distinct west–east gradient in economic performance, being lowest in the east. These areas have been established in the south-east region's planning policy making as separate subregions, comprising Kent, the south-central, south-western and north-western parts of the region. Essentially, the differences are between too little economic development and too high development pressures, reflected in a broad range in unemployment figures of between one percent and nine percent (www.gose.gov.uk/overview_f/index.html).

In addition to these acknowledged economic subregions, there is another set of policy territories layered on top of the subregions. The Learning and Skills Councils (LSC) and local branches of the Small Business Service (SBS), and the areas defined under the EU Structural Fund criteria, such as the Isle of Thanet around Ramsgate at the south-eastern end of Kent, present further subdivisions. The LSCs (formerly Training and Enterprise Councils) comprise employers, local authorities and other actors in local economies, and provide training and support and advice on training programmes to local employers. The boundaries of the new policies add further institutional responsibilities with their own territoriality. The arbitrary boundaries of these agencies have caused a realignment within London of local authorities, regeneration companies and business institutions.

Given the dynamism of the London region's economy, there are substantial uncertainties in the new economic governance and subregional arrangements, which may easily be overtaken by economic development processes. It is important, therefore, to have sufficient scope for institutional responsiveness to economic (re-)territorialization, which requires region-based policy-making capacity, including cooperative, informal linkages, so that institutionalized boundaries can be transcended in the interest of effective policy responses. At present, the new regional boundaries seem to obstruct such flexible approaches and subregional cooperation, in particular across formal regional boundaries, is merely tentative or non-existent. One example of the difficulties of horizontal and vertical cooperation and coordination is the Thames Gateway, following the Thames estuary and reaching from east London into Kent and Essex (Map 6.1). This development corridor thus not only encompasses several local authorities, but also all three regions – London, Eastern and South East. The inevitable result seems a plethora of responsibilities, competing policies and counter-effective initiatives.

Rolling back devolution – the Thames Gateway

The Thames Gateway wedge illustrates well some of the coordination issues across the new territorial and institutional boundaries of the three south-eastern regions and their RDAs. The definition of this substantial regeneration area (4000 ha) originates with Michael Heseltine's East Thames Corridor in 1991 which stretched from Stratford and the Royal Docks to Dartford, Gravesend and the Medway towns. The Thames Gateway covers extensive areas of derelict brownfield land and is cut through by the Channel Tunnel Rail Link (CTRL) with two new stations. LPAC identified 40 per cent of its strategic sites for manufacturing and warehousing growth in the area around the Bluewater shopping centre development and the river crossings of Dartford Tunnel and QEII Bridge. The Thames Gateway gains formal approval through its own Supplementary Guidance (RPG9a) outlining relevant regional planning policy. It is now considered a 'national and regional priority' (www.go-se.gov.uk/infor_f/thames.html).

The objectives of the Thames Gateway make more sense than those of the individual RDAs in that they refer to London as the centre of the region. Specifically, one of its objectives rests in 'enhancing London's position as a major World and European city' (RPG9a). The main focus is on re-development of old industrial sites (London area) and expansion of housing and business development, encompassing waterfront locations. Included are also the Medway estuary and other Thames estuary rivers. Both the Medway and the Essex Thames-side are also considered for port functions. The development is seen as a long-term, market-led process stretching over the next 20–30 years. A range of existing government programmes offers some financial incentives, e.g. through Assisted Area status, from English Partnerships and the Single Regeneration Budget. The changes in regional structures have divided the area across the three new regions into three sections. Below regional level are four district councils, a unitary authority, two counties and the London boroughs.

Subregional partnerships thus appear to proliferate. The borough councils of Dartford and Gravesham formed a joint executive committee in 1997. The Thames Gateway London Partnership (1995) joins 12 local authorities and the local training agencies. New public–private partnerships have emerged around potential development areas in Kent, in particular around the CTRL station at Ebbsfleet. The Kent Thames-side Development Agency links the major landowner, Blue Circle, with the county and district councils and Greenwich University. Subregional partnerships add to a complex pattern of economic governance. These bottom-up initiatives have been joined by three new development agencies. Three subregional groupings of public–private partnerships – East London, North Kent and South Essex – now add to the institutional landscape. The Thames Gateway is recognized in national policy and thus enjoys official, institutionalized status. It is thus perhaps not surprising that the Minister for the regions stepped in to resolve potential conflicts. The Thames Gateway Strategic Partnership (DTLR, 2000), which includes the chairmen and chief executives of the RDAs, takes a pan-RDA view, but, importantly, is directly under the Minister's authority. This demonstrates the continued involvement by central government in directly steering regional issues, despite theoretical devolution of responsibilities for economic development to the RDAs. Thus, central government continues to hold the levers of control, despite claims of facilitating devolution, probably in recognition of the difficulties established with the many territorial subdivisions of the economic area of the wider London region. Government faces a dilemma of control through funding and comprehensive planning guidance, and at the same time balancing central control with the encouragement of competitive attitudes within and between subregions, which is argued to underlie economic success.

London's regions – cooperation and the central state

The fragmentation of planning and economic development responsibilities poses problems at many levels. For central government, the search for effective governance at the regional scale engages processes of both deconcentration and devolution. The London region has three GORs. The effectiveness of this deconcentrated level of government has become a concern, leading to the Cabinet Office Regional Coordination Unit (Cabinet Office, 2000) being set up among other things to coordinate government departmental initiatives in the regions and to centralize regional management. The desire to get the deconcentrated management right and impose strong leadership from the centre may outweigh the potential for further devolution to regional agencies.

From the centre, regions are clearly seen as drivers of economic growth. The RDAs have to 'punch their weight' (DETR, 1997), putting them under pressure to perform. Inevitably, neighbouring regions may become rivals. London complains about paying too much tax and subsidizing other regions (Corporation of London, 2000). The three London region RDAs sit in very different political contexts, with coordination and cooperation mechanisms not being obvious. Does the LDA's democratic context give it a stronger voice? In contrast, the south-east appears weak due to its geography, lack of regional identity and the differing commitment of its counties. Kent County Council may see the RDA as a source of funds and policy influence, but other counties are less interested and view the regions merely as a drain on their resources. Party political control also varies across the region. The East Region Development Agency is conscious of its artificial boundaries (between Hertfordshire and Bedfordshire) that make cooperation seem essential but less easy in terms of administration and budget lines. Similarly, divisions are latent between the less well off coastal areas of Kent and the more affluent parts of Surrey and Berkshire. At the smaller scale, the LDA controls the SRB budget, and thus some coordination of regeneration and strategic objectives is possible. It does not, however, control the LSCs and SBS, so that coordinated subregional economic governance depends on the strength of local public–private relationships.

While government still holds the purse strings, the resolution of the fragmentation issues relies upon the initiative of local and subregional groupings.

With its financial control, however, the onus is on government to use that influence to facilitate its own policy objectives for regionalization. While the county councils and the mayor have strong planning roles, regional planning and arbitration are ultimately the roles of central government. There is, however, a long tradition through SERPLAN and LPAC of city-regional cooperation on planning issues. But regional and strategic planning is now split three ways, providing opportunities for disagreements and non-cooperation. New mechanisms will be needed to manage disputes, such as those about housing allocations or economic development or transport infrastructure.

London's world city objectives emerged through cooperative, cross-sector working during the 1990s without formal leadership and institutionalized government. Continuity of these relationships will have an impact on the development of cooperation in the present institutional context. The LDA brings in new actors from outside the LDP network. More diffuse networks may increase the time it takes to find consensus and cooperative approaches between LDA and subregions and individual players. Inter-RDA relations are entirely new. Unequal regional partners may struggle to find common interests as they compete for central government funding or inward investment as a benchmark of success. There seem to be few political or financial incentives to cooperate. In the Thames Gateway, coordination comes from the centre. Likewise, central government manages the RDAs through tight definition of their scope and budgets, while establishing a competitive culture of performance. In planning, the legacy of SERPLAN of cooperative informal arrangements to overcome the absence of institutionalized principles, is in danger, inevitably giving central government a stronger role.

The extent to which actors may wish to cooperate depends on their perception of economic interest and political expediency. What holds the London region together is the penetration at regional and subregional levels of central government. The government inserts new institutions, controls budgets and can roll out, or back, devolution where seen necessary. This includes changing the balance of political and financial incentives for cooperation between regional bodies. However, the current institutional reforms in the London region have, for the present, created cooperation and coordination problems for regional bodies that obstruct the search for an effective regional scale. The emphasis on competitive marketing and bidding for inward investment as the main rationale for regionalization in England thus seems to favour non-cooperation. In the London case, the new institutions may effectively undermine existing non-formalized cooperation which was borne out of the absence of formal mechanisms of regional government. It would therefore seem that two different models of cooperation may develop in parallel: (1) formalized, predictable and less threatening planning-based arrangements, and (2) competitive, essentially regionalist, economic policy-based structures leading at best to informal (and thus less committal) arrangements. The crucial question concerns the relative importance of the two models. In the London region, central government's focus on competitive bidding for investment has clearly pushed the balance in favour of informal, policy-centred, inherently anti-cooperative initiatives. With three regions covering the wider London area, it is thus likely that London will be divided into corresponding spheres of interest, challenging the government to act as mediator and facilitator of regional cooperation, rather than pushing towards further devolution.

The London case demonstrates the complex interaction of national institutional factors in the development of effective regional governance. World city London and an economically competitive south-east may be desirable objectives of government, but sorting out appropriate institutions is proving difficult. The core city clearly dominates. The penetration of the London economy into the wider region raises regional development and management issues, but also demands administrative boundaries that reflect economic reality rather than administrative convenience. The origins of the present set of institutions are influenced by the historical development of central-local government relationships. In London, we saw the impact of historical mistrust between central and subcentral government and the consequent incomplete devolution to mayoral government. The informal cooperation that pulled London and the south-east together, through LPAC and SERPLAN, has been undermined. In London and the wider region we also saw the continuing evolution of a particularly English style of public–private sector relationships. Business leadership has been encouraged, both at region-wide and subregional scale. But strategic coordination remains weak and government has been unable to guarantee the infrastructure for regional development. Basic infrastructure has been slow to emerge in the Thames Gateway, and upgrading the London Underground substantially delayed. The London case reveals the inherent tension between a desire for better regional economic performance and central control. Regional agencies are under close central supervision and regional investment too depends on the centre. Europe's unitary states may be moving towards a new regionalism, but in England's core city region progress in this direction is slow.

Monocentric regionalization under federalism: Berlin, Brandenburg and institutional obstructionism

Since 1989, and the end of the Wall, the relationship between Berlin and the surrounding Brandenburg state has been consistently on the agenda. This is not only caused by a seeming return to normality, i.e. allowing free functional exchanges between the core city and its region, but also the fact that administrative-governmental territorial separation remains. For the western part, the surrounding Wall (re-)translated into an administrative, if not physical, border. Examples of other such administratively divided city regions exist with the city states of Hamburg and Bremen (e.g. Baumheier,

1997; Huebner 1995). Their 40 years' experience in arranging – with varying success – cooperation between two *Land* governments demonstrated the difficulties of such arrangements, and it seemed opportune to – for historic reasons – consider a rapid move towards merging the two separate entities. This led to the failed attempt of merger in 1996 (Brenner and Heeg, 1998; Heeg, 2001, ch. 6). Typical city-regional development issues, such as have become more prominent in western German conurbations (Deutscher Städtetag, 1995/6), include the imbalances between development pressures and their costs in terms of public finance and environmental quality. Examples are the suburbanization of housing and business to edge-of-town greenfield sites. The main challenges are a growing gap in the territorial distribution of development and policy responsibilities. The rapid shift of the population into the (immediate) hinterland around Berlin shows signs of 'catching up' with the structural patterns of western post-industrial cities (SENSUT, 1996). Administrative boundaries play little role in such movements, unless they express themselves as relevant financial advantages, or disadvantages. Berlin sees region-wide cooperation as crucial for bolstering the city region's international competitiveness (SENSUT, 1996), not least in its ambition to join the world city league. Such collaboration is made difficult by the involvement of many formal players within the government system, from central to sub-local government roles, leading to a quite asymmetrical relationship between the two states (see Figure 6.2). Despite these arguments, formal collaboration is limited to a joint regional planning body, established as consolation prize after the failed merger attempt in 1996. Otherwise, collaboration between neighbouring local authorities requires informal linkages, involving, in the Berlin case, always the Senate of Berlin in its dual role as *Land* government and unitary *local* government for the whole of the city. The 12 administrative subunits, the *Bezirke*, are little more than parishes with few independent decision-making powers. As a result, in practice, the 3.5 million population represented by the local government of Berlin (Senate) are pitched against any one of the mainly small local authorities surrounding the city, many with only a few hundred inhabitants. It is not surprising, therefore, to note a sense of threat from an overpowering Berlin, with little prospect of genuine partnership among equals, although purely statutory provisions may imply it. Typical possible issues of such collaboration include the control of greenfield developments, technical and social infrastructure, local transport and leisure facilities. Much of this collaboration can only be achieved in smaller subregional arrangements, because the particularistic territorial structure increases the number of players rapidly with the size of the region.

Overcoming these difficulties in the interest of integrated region-wide planning in the Berlin area is a major political task, as there is little local experience to go back to. Earlier responses at the beginning of the twentieth century bypassed this problem by simply incorporating surrounding communities into the Greater Berlin of 1920. This is the same strategy as

Berlin		Brandenburg
Berlin Government (Senate) incorporates all of the following:		1–3 are separate institutions
1. *Land* (state) level: Senate acting as central government		1. *Land* (state) level acting as central government
	Sphere of possible	
2. Regional level: combined with local/central govt. role of Senate	**interaction between and**	2. Regional and level at sub-*Land* scale: • *Land* Govt. Office Regions (*Regierungsbezirke*) • Ämter (association of local authorities)
3. Local level: • Senate acting as local government for all Berlin • *Bezirke*: local adiministrative areas financially dependent on Senate	**across government** **levels**	3. Local level • larger: *Kreise* (small counties, upper tier of local government) • smaller: *Kommunen* (lower tier of local government)

Figure 6.2 Layering of territorial government: asymmetric relationship between Berlin and Brandenburg.

applied to London in 1965 with the creation of Greater London. Early attempts of establishing informal planning coordination between the counties (*Kreise*) around Berlin proved to be unproductive, as they did not include Berlin, effectively representing merely an empty core doughnut. The formal separation between Berlin and Brandenburg continued right up to 1945, with separate regional plans for each. The Iron Curtain, although magnifying and bringing to an extreme the separateness between Berlin (West) and the region, could build on established lines of administrative divisions. The absence of a joint administrative history meant that after 1989 no earlier common governance model was available to revert back to. Joint cooperation and, especially, merger, will truly enter new ground. At the technocratic level, however, cooperation has been developed for individual issues, such as public transport and utilities. These have been addressed across the border, because it is there that technical necessities and mutual interests, such as cross-border commuting, facilitate agreement. The situation is more difficult in competitive issues, such as attracting business investment or new local tax-paying residents by offering greenfield sites. In 1989, faced with rapidly emerging development pressures, Berlin presented various regional development models, including sectors (along the commuter lines) and multi-nuclei settlement models. In 1992, a regional development concept, jointly commissioned by the two *Länder* of Berlin and Brandenburg, suggested an inner and outer development area for Brandenburg, with a ring of settlement nuclei as expanded existing towns (SENSUT, 1996). Importantly, none of these concepts included the *Land* and newly empowered local governments of Brandenburg, which, especially in the latter case, were at that time too inexperienced and unprepared for effective active participation. Berlin had thus clearly taken on the leadership role in the region as a natural process, something later resented by the newly empowered (small) local authorities in its hinterland. This underlying resentment and mistrust towards Berlin as the implicit centre of power has continued to affect inter-governmental relationships between the two *Länder* at all spatial scales and functional levels. The move is thus towards fewer arrangements with less governmental underpinning, such as the Regional Planning Associations (*Regionale Planungsgemeinschaften*). They are groupings of local authorities for the purpose of designing regional development plans, here for the five sector-shaped planning regions in Brandenburg, including Berlin as 'sixth' (see Map 6.2). The link with Berlin is established through informal meetings at a 'regional planning conference', comprising the heads of government both of Berlin and Brandenburg. Since 1996, the Joint Land Planning Body has provided a more formal, institutionalized framework for cooperation between Berlin and the immediate hinterland of the 'inner ring' of Brandenburg. No such links exist with the outer areas, although they are part of the same five Brandenburg planning regions (Map 6.2). The Body's main responsibilities comprise the design of Land Development Plans and Programmes, approval of Regional Plans, ensuring that local planning complies with *Land* planning objectives (as guidance), and management of supra-local planning procedures.

Having been cut off from Western European urban and economic development as a western outpost in socialist Eastern Europe for more than 40 years, Berlin was quite abruptly stripped of this special status and thrown into the market place of competing urban centres. Initially, boosted by its designation as the new German capital, expectations were very optimistic. High growth rates were predicted for the city region, acting as a strong growth pole for the wider, rurally structured northern half of former East Germany. Expectations, however, did not quite materialize, as fewer than expected investors came forward, and even fewer wanted to set up their HQ in the city (Heeg, 1998; Lenhardt, 1998). Effective regional governance could have bundled resources and avoided duplication of public services within the region, but the decision to separate Berlin from its wider hinterland, Brandenburg, by a *Land* boundary, made such cooperation difficult. Moreover, although for historic reasons the functional integration of the city of Berlin with its surrounding region has been less developed than in

Map 6.2 Regionalization in the Berlin–Brandenburg area.

other German city regions, the rapid increase in infrastructure investment essentially required a regional approach. An extended international airport, freight centres, rail and road routes, and big retail centres have had to be planned and/or built within just a decade, due to the previous neglect of public infrastructure.

This situation with all its many implications for daily government decisions suggests that regional government of the city region Berlin should be well developed. However, since, like many other city regions, e.g. London, Berlin is divided by jurisdictional boundaries, regional governance can only come about if cooperation between the different governments can be accomplished, and a sense of common interest and purpose can be established. The evidence from interviews with local and regional actors, government documents, papers and newspaper articles, however, suggests that cooperation is only achieved in few and limited policy areas, and there, concentrated in a ring immediately around the Berlin–Brandenburg boundary. Outside this 'inner ring of functional interdependency' (see Map 6.2), such collaboration is much less obvious, at least not in institutionalized form. Nevertheless, all decision-makers agree, on an abstract level, that a common region would be beneficial for both *Länder*, Berlin and Brandenburg. In reality, however, institutional arrangements effectively encourage competition rather than cooperation, especially when it comes to fiscal revenue as a prerequisite of policy-making autonomy. Berlin, like the city states of Hamburg and Bremen in north-west Germany, is quite special among Germany's *Länder* due to its dual nature as local and *Land* government. These correspond to Brandenburg's two tier division into *Land* and local governments. Consequently, Berlin and Brandenburg communicate in a quite asymmetrical framework, because populous Berlin represents one local authority within city limits, with no institutional separation from its role as a city state. Any local authority seeking to communicate with its local partner, Berlin, will not know where the city's role as *Land* government begins. The uneven spatial scales of government between '*Land*' and 'local' are illustrated in Figure 6.2. Inevitably, this complicates matters, because possible interactions between the various levels of government institutions are not always straightforward, particularly between local government (*Kommunen* and *Kreise*) in Brandenburg and the city of Berlin wearing its *Land* government hat, caused by a sense of dealing with 'big brother'. The resulting complex mesh of interrelationships is illustrated in Figure 6.2.

Berlin and Brandenburg – monocentric and divided city region

There are great structural, economic and geographic differences between the city state of Berlin and its surrounding hinterland, Brandenburg (Table 6.1). Differences focus on the highly urbanized space economy and a largely rural, sparsely populated surrounding region. Table 6.1 illustrates some of the contrasts between the two *Länder*, which suggest those of a core and 'periphery' in a city region.

Berlin, with its 3.4 million inhabitants, is an island in the middle of Brandenburg, far away from the Western European economic centres. Due to their history, both East and West Berlin remained economically and administratively separated from their natural hinterland of the Brandenburg region. For West Berlin, surrounded by the Wall, any links to the GDR Umland had been impossible, and in East Berlin, suburbanization was politically unwanted (Benz and Koenig, 1995) as it was seen as a sign of bourgeois

Table 6.1 Some indicators of structural difference between Berlin and Brandenburg

	Inhabitants (millions)	Population density (per sq. km.)	GDP per capita (DM million)		Employed (000)		Administrative structure: no. of LAs
			1991	1996	1990	1995	
Berlin	3.6	3,900	35	44	1,667	1,484	1 (12 sub-areas)
Brandenburg	2.6	87	14	27	1,170	1,049	1300

Data source: MUNR/SENSUT (1998).

lifestyle and a weakening of East Berlin's presence as the GDR state capital. Most of the regional infrastructure circumvented West Berlin, establishing lengthy detours to the eastern half. It has taken until now, more than 10 years after the Wall, to reconnect the western half of Berlin with its hinterland. Administratively, however, albeit of a very different quality, the path of the old Wall continues its separating presence. The continued psychological impact of the former Wall gives the administrative boundaries between former West Berlin and the surrounding (formerly East German) Brandenburg quite a distinct quality compared with the similarly organized city states of Hamburg and Bremen. This is evident in today's relationship between Berlin and Brandenburg.

Being aware of the suburbanization problems in other (western) city regions, Berlin has tried to exercise some influence on the developments in surrounding Brandenburg right from the beginning of the post-Wall times, not least out of an understanding of being the natural leader in developments in the city region. Initiatives thus had a distinct Berlin-based perspective. This includes concern with the surrounding countryside as an area of recreation for the Berliners, especially those deprived of such access in western Berlin. Brandenburg was thus clearly perceived as the natural functional hinterland serving Berlin's needs. The scope for efficiency savings in service delivery, an important issue in the light of the soon-to-end subsidy payments to Berlin, was another driving force. Moreover, joining forces with Brandenburg was regarded as useful for a stronger bargaining position with other *Länder* and the federal government (Hassemer, 1995). So, as early as 1991, a commission was set up to explore the implication of a merger between the two states (Aigner and Miosga, 1994).

Not surprisingly, the separation between Berlin and Brandenburg has been perceived as 'unnatural'. Hence, the Brandenburg Prime Minister, Manfred Stolpe, described his state as 'the *Land* with a hole in the middle' (Cochrane and Jonas, 1999: 155). Its economic structure and development depend on Berlin, yet there is no administrative or governmental access to this economic base. This is of particular importance, as there are no other major metropolitan areas in Brandenburg, giving Berlin by far the dominant position in its urban hierarchy. There are only four cities exceeding 50,000 inhabitants, less than one-sixtieth of Berlin's size. Even Brandenburg's capital, Potsdam, can muster only 130,000 inhabitants, and it is an integral part of the Berlin conurbation. There is massive structural unevenness in the Berlin city region between the urban core and the rural, sparsely populated ring of Brandenburg. This is the case more so in the outer areas, with their much lower development potential than in the parts closer to Berlin. These discrepancies, and thus policy requirements, are not reflected in the administrative arrangements. In some areas, like in north-west Brandenburg, local authorities joined together in self-help development policies.

Given these economic dependencies and the expected cost of establishing a western standard infrastructure, Brandenburg's government was not hostile to Berlin's proposals to merge the two *Länder*. Both parties hoped that such a move would contribute to the economic growth of the region as a whole by providing a framework for collaboration rather than competition. Since the merger would have come into force after the period of office of the incumbent government, none of the politicians would have had to give up their positions. Their rationale was supported by projected savings of about one billion Deutschmarks through reducing administrative duplication and negotiation costs in the wake of a merger (DIW, 1995). These included jointly created and managed infrastructure, environmental management, water supply, etc., rendering obsolete an otherwise complex administrative jungle of more than 200 inter-*Land* contracts. Also, the joint population of some six million was deemed the necessary minimum for adequate administrative capacity of a *Land* government.

In short, there have been strong economic arguments in favour of region-wide administration and government through closer cooperation between, or even a merger of, the two *Länder*. Such cooperation, however, has faced considerable political and institutional obstacles as outlined below.

Obstacles to cooperation between Berlin and Brandenburg

The two main obstacles to formal cooperation or even merger have been political inheritance and institutional finance. The former revolved largely around the traditionally dominant role of Berlin as capital of the GDR and national capital before that (Benz and Koenig, 1995) vis-à-vis Brandenburg as *Land*. The underlying animosities and resentments, inherited from the days of the GDR and East–West divisions, have largely been held responsible for the popular rejection of the proposed merger between the two *Länder* of Berlin and Brandenburg during a referendum in 1996. This was despite support by the two governments and other organizations. Instead, issues of identity and political self-government were paramount, highlighting differences between 'merger-friendly' West Berliners and their more hostile eastern counterparts who feared a takeover by the West. More than half rejected the proposal. In Brandenburg, rejection was rooted in the old animosities towards a traditionally dominant Berlin as seat of government. The view emerged during interviews that the timing had been too ambitious, but that the merger is inevitable.

To rescue the situation and still achieve some form of desirable – from a government perspective – formalized cooperation, an annual joint cabinet meeting, and a biannual joint session of the two parliaments was agreed to facilitate an exchange of policy-relevant information and suggest a sense of good neighbourliness. Although a step forward in formalizing coordination, in effect, cooperation is more rhetoric, than actual policy making, with largely non-committal, formulaic statements of intent, e.g. 'progress was made in discussions aimed at future joint projects'. The only tangible joint initiative of relevance has been the institutionalization of collaboration in regional planning, the Joint Planning Body (*Gemeinsame Landesplanungsstelle*) in 1996. Its remit, however, is restricted to the immediate area around Berlin, the so-called 'inner ring' of strong functional interdependencies between Berlin and Brandenburg. For that area, a regional structure plan is to be developed, which will have no legal status and may be ignored by local planning authorities. This makes the whole arrangement something of a toothless tiger, and reflects the nature of the agreement as a compromise with no surrender of any statutory rights. This reflects a general storyline: any arrangements, which would mean 'costs' either financially or in terms of policy-making autonomy, have either been abandoned or only been accomplished after years of quarrels, and then only in a watered down version. For instance, the initial plan of a joint economic development agency was dropped amid financial squabbles as reported in the *Tagesspiegel* newspaper (30 Jan. 1998: 'IHK hält Länderfusion auf der politischen Tagesordnung' – Chamber of Commerce keeps *Land* merger on the political agenda). *Financial* cost seems to be the Achilles' heel of any potentially closer cooperation between the two *Länder*. This is largely due to the difficult fiscal conditions in Berlin since unification.

The causes of the financial problems of the city are diverse, but a predominant one was the rapid reduction of federal subsidies such as Berlin Aid once the borders between East and West Germany had opened. This meant a loss of half of the city's previous fiscal revenue. Between 1992 and 1994 most of the subsidies were phased out and, although some compensation was given by including Berlin into the inter-*Länder* fiscal equalization transfers, the city faced a shortfall of eight billion Deutschmarks. Further problems were caused by the decline of Berlin's old industrial structures, leading to a loss in business and personal income tax revenue. Furthermore, during the second half of the 1990s, Germany went through a general economic downturn which had an impact on nearly all revenue sources. At the same time expenditure rose for investment in the eastern part of the city and for 'unification projects' (DIW, 1997, 1999).

Competitive institutionalism between local and regional scale of government

Competition both at *Land* and local levels in inter-*Land* relationships is a major cause of coordination problems in the city region of Berlin and is further confounded by horizontal competitiveness at the local level of government, as well as, vertically, between government tiers (see Figure 6.2). An important reason is the distribution of responsibilities across the different tiers of government. Thus, many development-related issues within the city region fall into the exclusive remit of local authorities (Petzold, 1994; Schmidt-Eichstaedt, 1994), encouraging localist thinking despite the many likely external effects of development processes on neighbouring authorities. Some form of cooperation is thus essential, ultimately imposed by *Land* government planning guidance in the Land Development Plan. The extra complication in Berlin's case is that the city state combines both functions, thus effectively controlling itself. Berlin is thus by far the largest single local authority anywhere in Germany, and dominates the small, often rural local authorities in Brandenburg. And that is exactly what they are afraid of, and why they adopt a very cautious, approach.

Unsurprisingly, therefore, cooperation develops rather slowly, with localist 'isolationism' in policy making more prevalent, as revealed by a comprehensive survey in the mid-1990s (Bufalica, 1995). This revealed that little concern had been invested in cross-boundary cooperation. Where contacts between the authorities existed by the mid-1990s, they were usually limited to those required by statutory administrative processes, such as technocratic procedures during planning processes. There was little evidence of a likely change in attitudes, despite a general recognition of the growing mutual interdependencies. Instead, local concerns and interests were emphasized. There are few signs of fundamental changes to that. For instance, the four neighbouring boroughs (*Bezirke*) of the strategic area *Stadtraum Ost* in former East Berlin worked in parallel mode with no intercommunication. These findings confirm observations made regarding localist tendencies across eastern Germany (Herrschel, 2000). The main reasons turned out to be the lack of experience with the new system, and engrained, historically evolved separatist views between the Berlin and Brandenburg authorities. They were, therefore, not just a result of the Wall. The following section gives an overview of the major institutions of local government in Berlin and Brandenburg. On this basis it is shown that the observed reluctance of local authorities to cooperate is a predictable result of rational, self-interested behaviour within the framework of these institutions.

Local government throughout Germany commands a high degree of fiscal and policy-making autonomy, but much of actual policy-making capacity depends essentially on population sizes and the location of successful businesses, because they determine the size of the local tax base. This arrangement essentially pits local authorities against each other in the inevitably ensuing competition for attracting a larger tax base. An alternative option might be collaboration to maximize the likely success of such attempts, but there is little evidence of such an option being adopted. This points to a potentially quite limited local institutional capacity to exercise the so highly valued policy-making autonomy. Such finance-based competition is not evident in Britain, where local finance is controlled almost entirely by central government. In Brandenburg, the territorial organization of local government is very small in size, with a large number of competing players. In 1996, some 60 per cent of the then almost 1,700 *Kommunen* (districts) had fewer than 500 inhabitants, and only five per cent had more than 5,000. Although some of the smallest local authorities have merged since the mid-1990s, still almost 60 per cent of them have fewer than 500 inhabitants (Landesamt für Statistik, Brandenburg, 1999), reflecting the general lack of urban centres in Brandenburg. This fundamentally affects their capacity, and confidence, to transgress narrow interpretations and applications of statutory provisions and seek additional support for local governance, e.g. through cooperation with other local, or regional, actors.

The Brandenburg *Land* government has sought to counteract this problem by coercing the small authorities into forming administrative associations with no governmental functions, the 158 *Ämter*, to pool some administrative tasks. The rather limited powers of these *Ämter*, adds little to local government policy-making capacities, where particularistic, almost atomized territoriality makes collaboration seem an inevitability to achieve effective governance. In reality, however, such considerations seem to matter little. Against this backdrop, the *Land*'s own development plans (*Landesplan*) seek to provide a more comprehensive strategic framework as effective local planning guidance. The *Land* plan identifies five sectorally shaped planning regions, stretching outwardly from the edge of Berlin, and being dissected by an inner ring of high growth dynamics and an outer, peripheral ring of much less potential. The former is expected to be the main area of suburbanization through overspill from Berlin, and consists of 276 mainly small local authorities, pointing to the large number of decision makers involved in Berlin's city-regional affairs. This diversity faces divisions on the Berlin side as well. Berlin consists of two tiers of local administration, the upper tier is constitutional local government taken care of entirely by the *Senat* which also acts as *Land* government. The lower level of local government, the 12 *Bezirke*, are boroughs without their own financial income. They receive their operating budgets from the Senate and are thus cushioned from the tax revenue effects of changing population or business numbers. This reduces the pressure to compete and maximize tax revenue and may well affect their planning decisions and perceived need to cooperate with their Brandenburg neighbours who have a much stronger such interest. This arrangement complicates intergovernmental relationships, because Brandenburg's local authorities have to deal with both the neighbouring *Bezirke of Berlin* and the *Senat* as *éminence grise*. This perceived imbalance in power reinforces the resentments held by the Brandenburg local authorities towards an overtly mighty, even bullying Berlin.

Administrative-constitutional obstacles to cooperation

The constitutional arrangements for local government policy-making autonomy focus policy makers' attention on day-to-day rather than longer-term strategic decisions, aiming at achieving instant, politically valuable, rewards for their policies, especially higher revenue and job creation. Sharing any expected 'success' with neighbouring authorities appears unattractive, even though it may improve the success rate. Politicians seek to follow public opinion which has little awareness and understanding of artificially constructed 'regions' as opposed to the locality. In any case, the complex constitutional arrangements make alliances with other decision makers difficult to negotiate, *inter alia* because of uneven territorial powers within and without Berlin. This contributes to the still latent animosities between Brandenburg and Berlin, with the Brandenburg neighbours feeling disadvantaged in negotiations with 'big brother' Berlin Senate as their direct opposite. The Brandenburg *Land* government believed that cooperation could come naturally as part of local authorities getting used to their roles. Such a top-down managerialist view appears to be characteristic of Brandenburg's policy understanding, as illustrated by its rather dogmatic approach to regional planning concepts. Instead, many (smaller) local authorities opted for less 'threatening' and comprehensive arrangements and created *Zweckverbände* (single-purpose associations) instead, or they began to look for non-institutionalized forms of collaboration (Lieberda, 1996). On the Berlin side, party politics appears to be very important, circumscribing possible or likely alliances (Economic Development Agency (*Wirtschaftsförderung*) Berlin, July 1999; Strom, 1996). The Berlin Senate has sought to formalize such arrangements by creating cross-boundary working parties of local authorities in the four sectors sub-regional of Berlin, north, south, east and west. Three to four annual meetings are envisaged, and there is some financial support to incentivize such initiatives by the Berlin Senate.

Fiscal obstacles to regional cooperation

Population numbers and local businesses are the main basis of local revenue. Thus, just over 40 per cent of the average local fiscal income is generated by central government grants (block grants) which are based almost solely on the resident population. Local taxes, primarily business tax and local income tax, are the second major source of income, and circumscribe effective local policy-making autonomy. The proportionate importance of the two main local taxes depends on the relative attractiveness of a locality to business. For those in less competitive locations, seeking to become a commuter village may well be a financially more rewarding (and, possibly, more realistic) policy goal than trying to attract substantial new business. Given this somewhat complex situation, it is not surprising that *Kommunen* attempt to go it alone when seeking to attract investors, even though some may realize that concerted efforts with neighbouring authorities may well have their merit. This may include collaboration between two economically strong local authorities for a bigger, more efficient and visible site and infrastructure development programme than each of them could have afforded on their own. Less enthusiasm may exist, in attractive locations, to being burdened with a neighbouring, less appealing, locality. The only firm basis of cooperation between local authorities is the building of sewerage systems, because they are highly subsidized by the *Land* government. Berlin's *Bezirke* are also increasingly beginning to compete for investors, but so far, primarily for combating the social costs of high unemployment and to bask in the image-enhancing appeal of an attractive retail presence.

Examples of non-cooperation

The battlegrounds which emerged during the past years directly reflect the diverging and competing financial and political concerns of the two *Länder*, as illustrated by competition for investors and a considerable use of resources to outdo each other. In the year 2000, the Brandenburg government agreed to cooperate with Berlin in developing the whole city region as one entity. Indeed, Brandenburg uses the Brandenburg Gate in Berlin as its symbol for marketing itself as the 'capital city region'. The same plan had already been announced in 1995, but never got past the agreement stage. Each side kept blaming the other for the lack of progress. Berlin and Brandenburg accused each other of non-cooperation, and, indeed, poaching businesses from each other. Thus, Berlin is reported to have lost several hundred firms to Brandenburg, but is fighting back through attractive financial packages.[1] This will encourage further subsidy warfare.

Self-help cooperation in the outer parts of the Berlin city region: informal

The experience of the planning region Prignitz-Oberhavel exemplifies the internal division of the wider city region, pitching the local authorities closer to Berlin, i.e. in the 'inner ring', against those further afield. Prignitz-Oberhavel in north-west Brandenburg is one of five planning regions in the *Land* of Brandenburg which extends from the outskirts of Berlin some 150 km into an economic periphery. There are, thus, considerable differences in growth potential at a local scale within the city region. The *Land* is dissected into sector-shaped planning regions reaching from Berlin to the edges of Brandenburg. Within each planning region, individual local authorities seek to maximize the benefits of being situated closer to Berlin, regardless of the situation of their cousins in the more distant parts of the same region. There are few signs of sharing among the 'haves' and 'have nots' in terms of development prospects. Feeling left behind, a sense of shared grief among the local authorities of the outer area, including Prignitz, has encouraged informal arrangements in the *Städtenetz Prignitz* in the outer part of the planning region of Prignitz-Oberhavel.

1 *Berliner Morgenpost* (10 Oct. 1997): Berlins Kampf um Firmen und Arbeitsplätze (Berlin's fight to keep businesses and jobs).

The Prignitz network is one of a handful of examples of best practice in flexible and innovative regionalization supported by the federal competition 'Regions for the Future' (BBR, 1999a; BFLR, 1997) within eastern Germany, with 26 throughout the country (BBR, 1998, 1999b). The 'Prignitz grouping' includes seven small to medium-sized towns (3,000–25,000), and is based on the traditional, cultural-geographical region of Prignitz. This, it is believed (from interviews with local and regional policy makers in 1998/2000) is a name which will be more readily recognized by outside investors, or tourists, than the individual names of the participating localities. The stated main objective of this alliance (BBR, 1999b), set up in 1995, is to maximize the use of indigenous potential as the basis of sustainable development. This includes developing environmental quality, new technology, and more specific factors such as traditional cultural regional identity and entrepreneurial skill potential (ARP Prignitz-Oberhavel, 1997). The *Land* con-tributes financially through regular funding of development projects promoted by the grouping, but does not explicitly finance this informal region. It is not, therefore, an incentivized region, but rather one based on perceived real commonality.

A second urban network, the 'North-West Brandenburg Initiative', consists of local authorities along a railway line to Berlin. This rail link has just been upgraded into a rapid transit line to attract suburbanization from Berlin. The network brings together local authorities across the planning region, i.e. the immediate Berlin agglomeration and the rural areas further afield. The *Land* is using the existence of the Initiative as a justification of its regional policy of imposing an integration of 'strong' and 'weak' areas (ARP Prignitz-Oberhavel, 1997) by placing them into the same planning region. But there has been no financial support to facilitate such coming together. The Initiative is entirely voluntary, based on collaboration between the six (more recently, seven) participating local authorities. The Initiative is seeking a higher public profile as a more formal city network by publishing a quarterly report. With the rail link completed, the Initiative is now shifting its attention to facilitating projects of sustainable economic development along this line. Some of the local authorities belong to both the Prignitz and 'North-West Brandenburg Initiative' and are thus effectively members of three overlapping regions, each with its own 'milieu' (Danielzyk, 1998): the official formal planning region, and the two informal regions based on interlocal collaborative networks in policy making. Despite such initiatives, the *Land* government maintains a distinctly centrally directed, ideology-driven, form of regional regulation and territorialization, allowing only limited local input into regional planning strategies. The officially declared approach of 'decentralized concentration' is to be implemented as *Land* policy (see e.g. Heeg, 2001, ch. 6; Schulte, 2000) based on the paradigm of a polycentric system (Albers, 1998).

Alternative forms of informal regionalization, driven by a pragmatic assessment of economic realities, are thus emerging as self-help solutions, suggesting that the official formal policies instigated by the *Land* do not quite manage to establish appropriate regionalization as seen from within the formal planning region. The self-help solutions are based on identified common interests, or problems, between neighbouring local authorities. The development of a range of informal, network-based regions, with sufficient policy-making effectiveness is institutionally possible. The new model is meant to overcome deficiencies identified in the existing formal provisions for regionalization and the gaps it leaves in local–regional communication. Thus, local planning and development interests are to take a stronger role in shaping regional objectives, and are expected to be more inclusive of the various regional actors and their strategic outlooks for regional development. Nevertheless, despite guidance by the *Land* to initiate new, interlocal co-operation as the basis of regionalization, there has been no real departure so far from the continued reliance on established institutional structures and practices, which may include working with entirely artificial, centrally defined formal planning regions (TLG, 1998). Externally imposed forms of region-alization, however, may face inherent problems of credibility among the main players and may result in less identification with, and enthusiasm for, specifically regional issues.

Scope for developing regional governance in monocentric city regions – the case of Berlin–Brandenburg

The case of Berlin has illustrated the fundamental difficulties of regional governance when based around strong local government. The institutional-ized focus on locally based policy making, including fiscal dependence on local economic performance and population counts, in the interest of greater autonomy in local decision making, encourages localist views in policy making. In the Berlin case, such competitive and essentially anti-cooperative thinking appears to embrace particularly those local authorities with fewer development opportunities and thus scope for developing and implementing 'voter friendly' policies. Such thinking also seems to influence policies at *Land* level, following similar institutional interconnectedness between economic and population growth, fiscal revenue and thus democratic policy-making capacity. Any form of cooperation will inevitably have to address the finan-cial implications, especially sharing costs in terms of both possibly foregone revenue-generating opportunities by allowing a tax-generating business to locate in a neighbouring authority, and sharing direct expenditure on joint projects, such as infrastructure measures, to improve access to new housing or business development on the neighbouring authority's territory. In the absence of some form of cost and income sharing arrangement, the immed-iate question at the beginning of any cooperation project will be 'what's in it for me?' in terms of both votes and budget. A shift towards cooperative localism seems to require facilitating institutional arrangements to address these issues, but also a distinctly long-term view of potential benefits of any such cooperation, and this is not easy to maintain in an election and policy success-driven four yearly time frame. Any cooperation would have to succeed in a cost-benefit assessment, evaluating, in particular, the implications of cooperation on local political and fiscal capital.

Monocentric city regions in unitary and federal systems – competitiveness or cooperation between core city and region

The examples of London and Berlin regions highlight the complex interrelationship between subregions in each city region. This includes the relationships between the urban core and the wider hinterland region. In both cases, the crucial importance of institutional and functional territoriality became apparent. These issues are much less circumscribed by constitutional provisions per se, than by established, or emerging, practices among policy makers at a particular time. Similar problems emerged in London as in Berlin, with the territorial separation of the cities from their regions. In both cases, central government retained a not inconsiderable influence on developments. Another similarity is the duality of formal, planning-based and informal, policy-focused initiatives. The difference affects the relative authority of the city compared with the surrounding hinterland, which may be circumscribed by formal or informal arrangements alike. It does not come as a surprise that sheer institutionalization adds to complexity and thus more difficult bargaining processes. In both cases, more institutionalization seems to be viewed as the automatic solution of city-regional governance problems, but the reality has been a growing fragmentation and interinstitutional competition. It is central government that needs to push for greater simplicity of mechanism for integrated city-regional governance.

7 Polycentric city regions
Between competitive localism and 'marriages of convenience'

Following the discussions of two of the main capital city regions in Europe, this chapter looks at the nature and workings of regions with a less obvious urban centre, because several cities compete for a region-shaping role. Conditions for regionalization, especially from the local level up, are thus quite different. The two regions presented here are Yorkshire and the Humber in northern England, and Saxony-Anhalt in eastern Germany. Both are positioned immediately at subnational level, and illustrate the processes of regionalization in two polycentric regions. Each contains a number of competing cities of comparable size and importance, and roughly even distribution, thus placing them in a Christallerian least competitive position. In contrast to the monocentric London and Berlin regions discussed in the previous chapter, polycentric regions have no single urban focus as linchpin of regional commonality, because there is no overarching hegemonic interest by the main city. Beside this structural similarity, they differ quite significantly in their origins, constitutional backgrounds and thus operational capacity, reflecting their respective national institutional contexts. While the Yorkshire and the Humber region has its origin in economic growth management of the 1960s, and is thus little more than the territorial container for the implementation of nationally defined policies, Saxony-Anhalt is one of the 16 states comprising the German federation. Thus, it possesses considerable governmental power, acting effectively as central government for the local level, as well as having a historic territorial tradition and identity. This contrasts with the much less powerful and much more recent creation in England, where economic geography played a role in the drawing of boundaries. As a result, there are quite different institutional and practical circumstances for any regionalization process, including the scope for internally, i.e. locally, defined regions. Nevertheless, as the two cases will illustrate, there is considerable interlocal competition which poses a challenge to any form of region building, and, at times, is more likely to facilitate affirmative localism rather than compliant cooperative regionalism. Furthermore, the conditions attached to funding available under EU regional policy and the Structural Funds provide a further important input into any region-building processes, especially in terms of territorial boundaries and interlocal cooperation. They effectively create subregional policy areas of project-oriented (single purpose) collaboration.

Yorkshire and the Humber region: localism and incentivized regionalization

The first of the two polycentric regions presented in this chapter, Yorkshire and the Humber, prior to 1999 called Yorkshire and Humberside, in reference to the then existing Humberside County Council as 'partner' of Yorkshire, is an artificial construct of the 1960s, just like the other old Standard Regions in England. They have now been relaunched as part of the devolution policies for England. The region consists of two main historic-cultural and geographic entities: first, the wider Humber River estuary in Lincolnshire in the east, with Kingston-upon-Hull as the main urban centre, surrounded (until 1996) by the county of Humberside. The abolition of Humberside County Council in 1996 has effectively removed one player at the regional scale and replaced it with four local units, the unitary authorities of the East Riding of Yorkshire, Kingston-upon-Hull, North-east Lincolnshire and North Lincolnshire. The second main part of the region is Yorkshire in the west, with the three main cities of Sheffield, Leeds and Bradford (Map 7.1). They represent the three distinct, historic subregions of South, North and West Yorkshire. These new divisions reflect the fact that 'Humberside never won the affections of its residents, and much of the population remained fiercely loyal to Yorkshire or Lincolnshire' (Gibbs *et al.*, 2001: 108). As a result, the region suffers from a distinct division between north and south of the Humber river, which 'remains one of the defining characteristics of the Sub-region's political and economic structure' (Gibbs *et al.*, 2001: 108), and points to the whole region's divisions. These include distinct 'uniqueness to industrial culture and institutional fabric' (ibid.) which reflect the region's subdivisions between Lincolnshire, the Humber estuary and Yorkshire. This division had found an administrative recognition, albeit abolished subsequently, in the West Yorkshire, South Yorkshire and Humberside County Councils.

The four main cities, Sheffield, Leeds, Bradford and Kingston-upon-Hull, each have about half a million inhabitants and are situated in sufficient proximity to each other to encourage competition for influence. The smallest distance, and thus the potentially greatest immediate competition, is between Leeds and Bradford. This historic competitiveness underpins a degree of localism which sits somewhat uneasily with requirements of, now proclaimed, regional collaboration. Yet, different economic specialisms, e.g. steel making in Sheffield and textiles in Bradford, traditionally provided alternative markets and thus less head-on competition than can be found now with the pursuit of essentially the same projected sources of business investment. The traditional economic specialization has contributed to the development of distinctive local identities and images, which radiate out into their

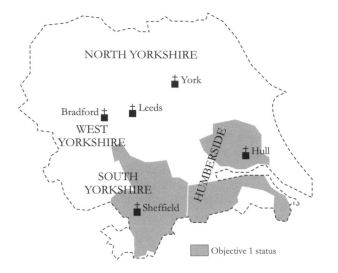

Map 7.1 Legacy and policy-based subdivisions in the Yorkshire and the Humber region.

immediate hinterlands. It is thus not surprising that the region of Yorkshire and the Humber seems more an amalgamation of smaller (and each in itself more coherent) subregions than a coherent regional entity. This is confirmed by the competitive attitudes between the three main cities in Yorkshire, especially Bradford and, immediately neighbouring, Leeds. Competition goes as far as ignoring each other's existence: Bradford seeks to ignore Leeds as a competitor (Bradford City Council, April 1999), while Leeds attempted to ignore the region altogether in its policy documents (e.g. Leeds Economic Development Strategy), and changed that policy only grudgingly after pressure from the Government. Effectively, it views itself as the dominant centre of the region (interview with LEDA, 17 September 1998). These divisions, and their localist perspectives, shape the ways in which the region functions. The divisions and effective boundaries are further engrained by the impact of the selective granting of funds under the EU Structural Fund Objective 2 policies (EC, 1999). Thus, the Humber and South Yorkshire subregions on their own qualify for support, not however the Y&H region as a whole. So it is in the interest of the local and regional players, supported by the Government through the Regional Office, for Y&H to highlight the, primarily economic, divisions. In terms of financial reward, so far, the Humber subregion feels less successful than its South Yorkshire neighbour. The de facto divisions are further supported by the Regional Development Agency (RDA) for Y&H, acknowledging four main subregions based on existing economic specificities and the underlying actor networks, representing stark variations in development prospects (Mawson,

1997a). These differences are largely represented by the four main cities in the region. This has translated into differential, even contrasting positions towards the region. Sheffield has traditionally seen itself as (the leading) part of the South Yorkshire economic subregion with a distinct territorial identity. This has now been underpinned by its Objective 2 status, together with the subregions of West Yorkshire and the Humber estuary. The old industrial conurbation of Leeds, by contrast, has shown much less of a regional affinity. As now, relatively, the best performing urban economy in the region, and thus not qualifying for EU funding, Leeds takes a distinctly locally centred, independently minded position. The exclusion from EU funding does not only confirm Leeds' insular situation in the regional economy, but also reaffirms its separation through a distinct, policy-based territorial boundary. This exclusion has not only removed one important reason for engaging in a regional debate with its neighbouring local authorities, but has also strengthened Leeds' sense of economic superiority vis-à-vis the region.

On that basis, Leeds is viewed as a regional subunit in its own right, together with West Yorkshire around Bradford, South Yorkshire centred on Sheffield, and the North Yorkshire Leeds–Harrogate–York triangle. Policies essentially operate at the smaller scale of these subregions, because they represent a more coherent identity and acceptance as areas of potentially shared interests, while the whole region of Yorkshire and the Humber is viewed as little more than a rather vague strategic policy container. This, together with a strong localism based on the main cities as foci of the subregions, will be illustrated here for the Leeds, Humber and Sheffield/South Yorkshire subregions.

Leeds: city without a region?

There seems to be a growing recognition even among localist city governments, such as Leeds, that development is going towards an increased role of city regions and thus, there is a need for cities, including sceptical ones like Leeds, to be aware of potential collaborators in the new arrangements (interview with LEDA, 17 September 1998). This recognition that intercity networks at city-regional level will be of increasing importance, finds some evidence in the fact, for instance, that even self-centred cities like Leeds are beginning to be drawn into a regional network. Thus, for instance, the chief executive of Leeds City Council has been chairing the Regional Assembly of Yorkshire and the Humber, and there is a sense that many issues can/will be better solved at regional than either local or national level (interview with LEDA, 17 September 1998). The problem, however, is that regions in England are difficult to define. There is a long way to go before localities simply join together in the pursuit of common interests and formulate a regional agenda. As a result, they don't develop any meaningful economic strategy (with enough local input). At the end of the 1990s, Leeds was suffering from a big city syndrome, believing itself to be independent of any region because of its successful economy (LEDA, 17 September 1998).

The competitive rather than collaborative attitude is illustrated by the Leeds Initiative, established in 1990 as a vehicle for more market-oriented, aggressive economic policy making. Here, 'the key institutions in Leeds came together to form a coalition, whose remit was to promote the city to the outside world' and to improve the perceived appearance of the city (Ward, 1997: 1500). This spatial coalition was distinctly local, with no explicit reference to, or inclusion of, the wider region. These new coalitions included actors not previously considered by the city government as part of the city's governance, as the reference by one councillor to them as 'partnership people' (Ward, 1997: 1500) suggests.

There is, thus, an underlying notion of a natural leadership for Leeds City Council as the elected centre of local government, and of Leeds within the region. Thus, one voluntary sector representative observed that 'Leeds has a very parochial view of doing things itself. It's partly cultural . . . Leeds doesn't play nationally . . . Leeds is a city state' (Ward, 1997: 1501). The last observation is in line with the impression given by the Leeds Initiative documents and their focus entirely on the city, as if in a spatial vacuum. So it does not come as a surprise that there are hardly any references to the region in Leeds policy documents. The idea of 'region' does not feature once among the main development objectives of the city's recent economic development strategy (LEDA, 1996). However, in response to the new RDA framework, and to comply with politically induced requirements, a regional focus has been written into the strategy (LEDA, 1998). In addition, councillors have recently become more involved in regional networks, such as the Regional Assembly. The city is attempting to work out a response to the new regional structures. At the same time, English regional reform has prompted Leeds to join other large English cities which form the Capital City Network, with the aim of influencing government policy from another, more city-focused, direction. Leeds has thus responded to regional reform by taking more interest in subregional issues, informal alliances and national networks. The distrust between Leeds and the wider region is mutual. Leeds is seen as vacuuming up all new investment, if necessary with the help of financial incentives, leaving little, if any, business investment for the rest of the region.

Subregionalization of Yorkshire and the Humber

The current Regional Economic Strategy (Y&H RDA, 1999) ('Yorkshire Forward') offers two 'themed' subregionalizations for Yorkshire and the Humber. The first, 'natural' one, is based on resources, i.e. a division into 'coalfield', 'national park' and 'other rural'. The second subdivision derives from externally defined regional policies on the basis of relative economic performance, in particular eligibility under the EU Structural Fund

(Objective 1 and Objective 2 status). These subregions are based on common characteristics and, thus, likely similarity in policy requirements. This similarity may arise from the need to tackle the common problems of economic restructuring, or from the politically important desire to maximize EU funding for local projects. It facilitates interlocal collaboration, if only for a limited time. Objective 1 status and the prospect of drawing down outside funding is thus one, if not the only, reason for these subregions' existence, established under the auspices of the GOYH and its European directorate. These divisions are explicitly acknowledged in the 1997 Economic Strategy as 'subregions' (GOYH Regional Development Strategy, para. 4.41), which also assumes policy needs aimed at specific subregional requirements. Each of these 'areas' is to be consulted about their regeneration priorities (para. 4.42) and serves as the 'essential first step to community led regeneration and will underpin all our programmes, thereby re-enforcing the Government's local democratic renewal agenda' (para. 4.43, p. 45). Another, if not dissimilar subdivision is being suggested by the Secretary of State's Regional Planning Guidance. Aimed explicitly at economic development, next to environmental and urban development policies, as the main rationale behind the whole regionalization programme, the Plan establishes three main spatial policies which effectively create subregional entities: the Humber Trade Zone, Deane Valley Development Zone and 'market towns as centres of rural enterprise' (p. 50).

This complexity at the regional scale, with varying layers of functional policy-based territorialization, is underscored by the fact that there are effectively three different planning and policy documents aimed concurrently at Yorkshire and the Humber's regional development, each setting out its own semiformal subregional divisions: (1) Yorkshire and the Humber RDA (trading as Yorkshire Forward) introduces its Regional Economic Strategy, (2) the Regional Chamber through the overall strategic regional framework 'Advancing Together', and (3) the Regional Assembly, as the body with the relatively most state institutional qualities, through Regional Planning Guidance and the Regional Transport Strategy. The RDA points out explicitly that 'subregional collaboration will provide the operational link between the Regional Economic Strategy and local delivery' (para. 4.64, p. 52), thus stressing the specific local, rather than general regional, benefits. This policy-territorial multitude corresponds with an array of new regional institutions operating within the region to differing agendas, with some creating their own subregional policy territories. The rapidly increased 'institutional thickness' established during the second half of the 1990s encompasses the Government Office (1994), Regional Assembly (1996), which was developed in response to local government initiatives and lobbying, the Regional Chamber (1998) and the Regional Development Agency (1999). All institutions, except the Regional Assembly, were installed and set in motion from above and point to the essentially top-down rationale and operationalization of the regionalization process in Yorkshire and the Humber, as it does for all English regions in general. The three documents' main common feature is their primary concern with economic development, and the marketing of the region to attract EU funding and inward investment. This follows central government objectives and policies.

Given these differential, fragmented, circumstances with central directives and interlocal competition, there are calls for a more locally based approach to regionalization, led by the cities as the main growth centres (such as envisaged by the Chief Executive of Bradford (Mawson, 1997b). The cities are to act as linchpins of regional awareness and policy making in, effectively, the individual city regions. Given the rather limited interlocal communication and willingness to cooperate between competing cities, such as Leeds and Bradford (interview with Bradford EDU, 1999), the outcome may be the building of new or higher walls between these city regions within the larger region. Indeed, official documentation by the cities hardly, if at all, refers to the region as a potentially supportive, or challenging, economic base. Certainly, it seems evident in this instance, that regions cannot just be imposed or installed. The case of Saxony-Anhalt in Germany, discussed below, confirms this conclusion. Effective and relevant policies appear to prefer locally defined territorial divisions of varying, problem-circumscribed, scale and with actual, practical, meaning. Understandably, there are different views of how a more bottom-up form of locally led regionalization might work. With little to gain from the wider region, Leeds considers itself a self-contained *city* region (LEDA, 1998), with distinct emphasis on *city* (interview 1999), all but ignoring any regional dimension beyond its immediate hinterland and certainly shaped by the core city's interests. Nevertheless, Leeds has been included in the emergent multilayered, multiscale arrangement of institutional governance which includes 'a number of quango-like agencies . . . either in Leeds or at the regional level' (Ward, 1997: 1500), which reinforce the introspective perspective taken by the city.

Sheffield – uneasy coexistence of city and its subregion

In Sheffield, by contrast, a more immediate and stronger regional outlook prevails, albeit focused on the immediate South Yorkshire subregion, for which the city claims 'natural' leadership, rather than the whole of Yorkshire and the Humber. This is partly for traditional economic reasons, with Sheffield still tied together with its immediate hinterland, especially in the north and east, based on the old linkages of the area's steel industry. Yet there is little contact with its neighbouring localities to the south and west (interview SCC, 24 September 1998), although new common interests may rest there. Possibilities include tourism in collaboration with localities to the west, in the Pennines. The preference of the 'trusted old' links, rather than exploring new possibilities, may be the sign of a lesser confidence in its role as a growth centre compared with Leeds. The latter seems to rely on its innate attraction as growth initiating partner for the rest of the region, providing it with new key roles in regional matters, without having to show any interest in it. Thus, Leeds obtained a stronger foothold in the new formal regional arrangements, e.g. chairmanship of the new Regional Assembly, than Sheffield, which has no such formal role, has managed so far (interview SCC, 24 September 1998). The city seems to enjoy less obvious visibility as a centre of investment and economic development (SRB, 24 September 1998), making it less coveted as a must have key actor for the region as a whole.

Much of South Yorkshire's appearance as a subregion is incentivized by the availability of EU funding for old industrial regions (interview SCC, 24 September 1998). Objective 2 funding focuses attention on the shared structural economic problems of the area and establishes a common interest. This manages to, at least temporarily, overcome localist antagonism between Sheffield and the smaller towns in South Yorkshire. Localist attitudes, rooted in Old Labour traditions (SCC, 1998) and strong unionism among the coal miners and steel, developed during the 1980s despite the history of cooperation in the South Yorkshire County Council, which in itself had taken an intro-verted, isolationist perspective. Since then, Sheffield has moved away from Old Labour politics towards a more entrepreneurial, business-oriented local policy, leading to a distinct contrast in political culture between its New Labour politics in the core of the city region and Old Labour values and principles in the South Yorkshire subregion outside the city. Nevertheless, South Yorkshire shows the importance of 'real' regional interconnections, based on economic and commuter links whose reach circumscribes the subregion. But strong localist tendencies in policy making persist, especially in the smaller localities. Economic policies there are largely led by traditional Old Labour paradigms, as in Sheffield in the 1980s. They still include protectionist views (SRB, 24 September 1998), and they obviously collide with Sheffield's new approach.

Although stigmatizing the area somewhat as underachieving, the extra funding is keenly awaited to support the cities' regeneration projects and, in Sheffield's case, its attempts to reinvent itself as a post-industrial city with pride in its still relevant industrial heritage. At the same time, the need for such an external stimulus to shape a subregional perspective points to the relative weakness in forming collaborative ties and an awareness of common interests in South Yorkshire. Interlocal alliances and policy approaches are seen as desirable only from a strictly pragmatic perspective. More flexibility and openness to overcome localist viewpoints among policy makers in South Yorkshire, shaped by legacies of competitive localism, both economically and politically, seem necessary. This, however, can only work if funding streams respond to emergent new regional links, and thus give them a visible advan-tage against non-cooperation. Without such bonus, the traditionally less developed regional identity and policy making in South Yorkshire (interview

SCC, 24 September 1998) would counteract any collaborative ambitions. Drawing down European funds has thus become the main regionalizing force in the sense of a 'marriage of convenience', based on a shared regional destiny and identity, at least for the time being. Thus, subregional cooperation and alliances may well change as new priorities and boundaries are set for the Structural Funds (Shutt and Colwell, 1997) or, indeed, as other funding sources appear. These shifts, rather than encouraging the building of new regional initiatives among existing networks, may strain existing alliances and the coherence of the wider region in the interest of the pursuit of new opportunities with partners elsewhere. But there is a danger of a 'them' and 'us' mentality between the successful South Yorkshire and the Humber subregions in attracting EU funds, and the other non-supported parts of Y&H (interview SCC, 24 September 1998).

At the same time, such obviously time-limited, instrumentalized views of regional interests and territorial and topical areas of collaboration may be the only form of cooperation feasible, because of limited commitment and collaboration 'by convenience'. No locality needs to feel it has to surrender any individual advantages in the interest of the subregion's common benefit. Not surprisingly, given the political-economic backdrop, the economic subregion of South Yorkshire offers only limited coherence in policy-making ambitions.

Political animosities, with accusations of ideological treason between South Yorkshire's towns and Sheffield, are reinforced by the anxiety that Sheffield is potentially striving to dominate the city region, effectively turning it into its dependent hinterland. This notion is supported by the resentment felt among the smaller towns and cities in the subregion about Sheffield seemingly scooping up all EU Objective 2 funds aimed at South Yorkshire. This is essentially the same accusation levelled (by Sheffield) against its rival, Leeds, if at a wider spatial scale. The relationship between city and its region is thus ideologically charged and potentially fraught (interview SCC, 24 September 1998). It remains to be seen to what extent financial incentives can smooth over this gap. Sheffield, it seems, is learning lessons in collaborative networking, such as identifying future possibilities in developing links with other urban localities for new policy objectives, such as establishing, and participating in, a 'tourism region' to the west and south-west of the city. More flexibility in regionalization is required. However, this can only be achieved if funding streams recognize the new regional links.

While such horizontal linkages are as yet undeveloped, there are strong vertical connections between city and national government, some with a somewhat fraught history. Sheffield is potentially emerging as the link between national and (sub)regional interests. These go back to the 1980s and the creation of the Sheffield Development Corporation (SDC) in the Lower Don Valley, when central government pressed on a reluctant Sheffield City Council (SCC) to accept an Urban Development Corporation. In the absence of alternative sources of funding, compliance with central government policies was the only available way to maintain at least some scope for implementing locally defined policies. Also, David Blunkett, as Council Leader in the early 1980s, is now a key member of the current Labour Government. In addition, the local Labour member of parliament, Richard Caborn, was Labour Party spokesman on regional affairs before the 1997 election and subsequently the minister with direct responsibility for regional institutional reforms. Viewed from London, and this perspective continues to be underpinning the whole process of regionalization, the economic performance and competitiveness of the regions has been the main concern. This economy-centred view has now been reinforced by moving responsibility for the RDAs to the Department of Trade and Industry. This adds to the complexity, because now two Government departments are responsible for regional issues. How far this affects the whole process and its purpose remains to be seen. From a local (Sheffield) perspective, the wish list for regionalization includes effective subsidiarity (because different issues are best addressed at different scales of government), a generally clearer framework for facilitating economic growth, i.e. a genuinely integrated and coherent strategy, appreciation by the city for the need to make a distinctive contribution to the region and thus engage more actively in its governance, and avoidance of counterproductive competition between different scales of government (SCC, interview 24 September 1998). This involves a clear division of responsibilities (interview GOYH, June 1998). By the end of the decade, several potential regional links offered themselves to Sheffield: the transport corridor (port) Sheffield–Hull for goods and tourists, albeit with currently no formalized links and territorial identity, the 'economic region' of Leeds–Hull, Nottingham–Derby, and as a potential, if as yet undeveloped, tourism region covering parts of the Pennines (interview SRB, 24 September 1998).

Divisions in the Humber subregion

The issue of localist divisions of interest have become particularly pronounced in the Humber subregion. Competitive striving for EU funding and, subsequently, the detailed conditions attached to any supported project, have been major contributing causes. Although this is not a problem unique to the Humber subregion, because similar issues have also arisen elsewhere, e.g. in South Yorkshire (see above), the somewhat diverse nature of the subregion has come to the fore. This has made the emergence of common interests and, subsequently, the agreement of a common strategic plan and policy focus all but impossible. In effect, 'given the disparate group of actors within the Subregion [of Humber] . . . , there was little commitment to a strategy for the Subregion as a whole, and individuals sought to obtain resources for their own local area' (Gibbs *et al.*, 2001: 112).

This emphasis on drawing down funding has also, if indirectly, increased the divisions between the subregions of Yorkshire and the Humber, because the Humber subregion continued to feel disadvantaged in favour of Yorkshire and its well documented and publicized economic problems and thus portrayed need for support. This increased a sense of political marginalization and disadvantage, and thus resentment (Gibbs *et al.*, 2001), contributing to competitiveness and determination in obtaining EU funding in whatever ways possible.

The Objective 2 status of the subregion not only effectively subdivided the Yorkshire and the Humber region overall, but also divided the subregions, such as the Humber. Thus, the part south of the river qualified for support as a whole, whereas in the structurally better off north, EU support was available only in Hull with its old port industries. The other parts of the region were excluded (see Map 7.1). In addition, some areas were also eligible for more specifically targeted community programmes for old industrial, steel and mining areas, such as RESIDER (Scunthorpe), KONVER (Hull) and PESCA (Hull, Grimsby). Facing confusion and division, a Programme Monitoring Committee was established to facilitate at least some coherence between the many policy initiatives (Gibbs *et al.*, 1999). The absence of territorial boundaries corresponding to the diverse nature of territorially targeted policies added to the sense of confusion (Gibbs *et al.*, 1999: 10). The EU policies thus effectively created a separate set of territorial policies, where the areas are no more than formalized spatial containers for policy implementation, and are incongruent with existing patterns of regionalization. For instance, some locally based, informally agreed initiatives had emerged prior to the Objective 2 status, 'promoted by pro-active local authority officers within some parts of the subregion'. This points to a perceived necessity of such subregional collaboration, driven by policy expediency in the pursuit of specific (funded) projects rather than a sense of genuine (sub)regional awareness. There is no indication of such moves for the Y&H region as a whole. The tentative arrangements between neighbouring groups of local authorities, reaching into the (sub)regional scale, however, are subjected to the essentially contradictory effects of the new EU funding status, encouraging both collaboration of local interest and a localist mentality of maximizing purely local financial benefits. Any collaboration is narrowly focused on particular projects submitted for funding, but there is little sense of a genuinely shared common regional agenda. The somewhat divisive impact of the EU policies is reflected in the Single Programming Document, developed by the Government Office (GOYH) in response to EU funding opportunities, and its successor, the RDA's Regional Economic Strategy. Their contents and policies are essentially a 'shopping list for EU funds' (interview GOYH, June 1998). Given the central importance of the EU Structural Fund, it may come somewhat as a surprise that a lack of experience by the European Secretariat at the GOYH has been criticized by ambitious local authorities (Gibbs *et al.*, 2001: 113).

Nevertheless, some attempt at establishing a degree of coherence to the fragmented landscape of policy interests was made, such as in the form of the Area Advisory Group (AAG) established to help coordinate submissions to EU funding rounds. But this Group 'lacked any strategic vision' (Gibbs *et al.*, 2001: 112), adding up to little more than a 'rubber stamp committee' (ibid.). Thus, 'the delivery of the Single Programming Document, [outlining the various projects in the regional context,] remained dominated by individual project-based initiatives' (Gibbs *et al.*, 2001: 114). As a result, initiatives lacked a coherent strategy and many overlapped or, worse, were counterproductive. There was little effort to establish and operationalize a regional agenda. Instead, there was an overriding concern with individual projects submitted for funding, and their implementation in compliance with funding terms, irrespective of their contribution to a wider picture and development of the area. The abolition of Humberside County Council as strategic body for the whole of the subregion, encouraged the creation of a policy discussion platform, the Humber Forum, in 1996. Its purpose was to bring together the various actors under a (vague) regional agenda. True to the meaning of its name, it offers an opportunity to meet and facilitate 'partnership acting on behalf of the subregion' (Gibbs *et al.*, 2001: 114). As it is voluntary, it is perhaps not surprising that there is little evidence of coherent planning and policy making at the (sub)regional level but rather, a plethora of plans at various spatial scales from local to region, involving a multitude of actors and as many different interests and policy objectives. Resulting contradictions and overlappings of policies are thus inevitable, and commitment to specific, particularly longer-term policies, very difficult. This points to a considerable 'weakness of partnership dynamics' (Gibbs *et al.*, 2001: 114) in the subregion and beyond, and raises questions about the integrity and effective operationability of the whole region when subjected to competing local and subregional interests.

Effectively, this fragmentation reinforces the central role of the GOYH as manager, instigator and common focus point of attempts to represent the region, if only to draw down EU funding. The Area Advisory Group (consisting of a range of subregional actors), intended to provide a strategy-shaping platform for the subregion, was seen as little more than a 'rubber-stamping committee' (Gibbs *et al.*, 1999: 12), similar to the situation with the other subregions' AAGs. This points to a continued, strong top-down influence by central government in regional matters, despite the public emphasis on regionalization (and devolution). Despite its shortcomings, the Humberside AAG was seen by some as a useful expression of the subregion's clearer identity within the larger region of Yorkshire and the Humber, not unlike South Yorkshire, than was evident for the other two subregions, West and North Yorkshire (Gibbs *et al.*, 1999: 13) without incentivized regionalization.

This situation changed little with the newly established RDA which institutionalizes a differently focused, business-led rather than planning oriented approach to the region, concentrating on formulaic strategies. By its nature, a more region-specific strategy could be expected to maximize the use of indigenous potential as the region's appeal to business. In effect, however, as in the other English regions, the new strategy was written in a hurry and in conformity with government guidelines (DETR, 1998a), and was driven by current political agendas, such as the Labour Party's Annual Conference in late 1998 (interview GOYH, 1998). Given the rather centralist, top-down approach to contents and implementation, it is not surprising that concern was registered in the region's main cities that the generalizing central government perspective in regional policy making would not be able to reflect intraregional differences in problems and development opportunities (Leeds DA, 1998; GOYH, 1998). This may reduce the perceived adequacy and, indeed, relevance of resulting regional strategies and advocated policy goals on the ground. Moreover, given central government's emphasis on maximizing the amount of money drawn down from the EU's Structural Fund, and the pressure on RDAs to perform, the emerging Regional Strategy appears to be little more than a government-directed and approved list of projects with the highest likelihood of attracting EU funding, rather than a strategy aimed at establishing and strengthening (indigenous) regional economic development capacity. This includes acknowledging the de facto subdivision into the four subregions of North, South, West Yorkshire and the Humber, each with their particular backdrops to regionalization both in terms of structure and policy making. Nevertheless, with EU funding needing to be channelled through the Government Office, there are tentative signs of a network relationship emerging across Yorkshire and the Humber, highlighting the role of the Government Office as a common reference point for the whole region. These 'grant networks' developed because 'organizations knew that if they did not work in partnership they undermined their chances of getting discretionary resources for economic development' (Bache, 2000: 587). The quality of these relationships consequently varies between largely cosmetic collaboration to maintain a stake in potential funding distribution, and more genuine, active networks, based on perceived potential general benefits of collaborative partnerships in the pursuit of identified common policy goals (see Bache, 2000). By its very nature, such regionalization is rather superficial, often effectively little more than a common label used by the relevant group of local authorities. This heterogeneity was mantained because of the diverse, and changing, foci of collaboration on individual projects submitted for funding. Thus, despite the emphasis on partnership building as part of the programming procedure under Objective 2, 'the delivery of the SPD in the Humber Subregion remained dominated by individual project-based initiatives, rather than broader strategic partnership working' with its wider, shared supra-local perspective. In effect, 'partnership was not a *sine qua non* at project level', but was limited to collusion at the specifics of individual local, if regionally argued, initiatives (Gibbs *et al.*, 1999: 14). The result has been an array of projects and associated layers of varying territorial entities and boundaries with little evidence of a coherent, indigenously supported regional under-standing. The emphasis of the 1994–1996 SPD on 'old industries' as economic base highlights the importance of available funding as a policy-shaping feature, here Objective 2 funding status, and this created a 'policy region'. Thus, the economic subregion of the Humber was referred to in the SPD as 'the county's economy', thus not only pointing at the lack in clarity of meaning, but also scale and rationale of the subregion.

The other main consideration leading to a positive view on the 'region', essentially also financial, is the perceived potential benefits of having a voice in Whitehall representing their interests, albeit in competition with the other English regions representatives in Whitehall. In effect, therefore, the GOYH has emerged more as a facilitator rather than a monitor of EU funding for the region, and is now pivotal to links and networks across the region (Bache, 2000).

Not surprisingly, 'there was little evidence of attempts to avoid overlap [between initiatives], either within the Subregion or across the region as a whole' (Gibbs *et al.*, 1999: 14), and seeking to 'integrate projects' (Gibbs *et al.*, 2001: 114). This may well be explained by the temporary, incidental and purely pragmatic nature of these boundaries of policy territories in response to project management rather than territorial meaning and strategy. The strong localist standing by the four main cities has added to this obscuring of regional issues by the more immediately pursued and accepted local inter-ests. Often, it was assumed, and confirmed in practice, that the GOYH would take that role of forging and operating the region (Gibbs *et al.*, 1999: 14), thus effectively presuming a governing from above rather than from within the region. In this relative vacuum of bottom-up scope for expressing regional interests, however, it was left to the AAGs to encourage and develop part-nership, cohesion and strategy, although the simple inclusion of several actors did not automatically lead to such coherence, but followed funding opportunities. The result has been a rather fragmented, opportunistic and incidental nature of regionalization, with individual (local) interests likely to prevail. Inevitably, the outcome may be a hotch potch of separate initiatives driven by the availability of funds (and funding conditions).

This essentially short-term, project-focused perspective of policies is the outcome of almost 20 years of growing centralization and disempowerment of the local, let alone regional, scale of governing. In addition, the empha-sis has been increasingly on fostering interlocal competition for available funding, rather than encouraging collaboration and communication. It remains to be seen to what extent the current changes, with a claimed return of the 'region', can change this pattern. It seems to be obvious, however, that region-alization per se cannot simply be imposed. 'Particularly in areas characterized by significant political and economic fragmentation, notions of "the region" cannot be taken as given' (Gibbs *et al.*, 2001: 116).

Summary

In the Yorkshire and the Humber region, currently, there is strong competition between spatial policy objectives at different scales. These translate into locally focused and truly regional policy making. Their contents and relative importance still depend on external factors, either in the form of central government strings and directives, or EU funding incentives. They seem to make regions of cooperation. Strong local identities and ideologies raise barriers to for such processes, and it will be the regional engagement of the four main cities that is likely to make or break the region's coherence and identity as a political-economic space. Existing, if differing, informal networks in the cities are seeking ways of responding to the many new regional institutions. Local development objectives, without much in the way of regional credentials may succeed because the new RDA needs to prove its effectiveness and good projects from the (main) cities will help achieve that aim, possibly at the expense of the smaller localities and a more integrated strategy for the region. Differentiation within the region and fragmented regional strategies are to some extent a distinctive feature of Yorkshire and the Humber, underlining its politically constructed nature. The importance and positive effects of an accepted regional identity become obvious from looking at the West Midlands region, for instance, where a strong regional identity creates considerable positive spin-off effects, such as joint lobbying and strategic policy making. In that case, however, we have an essentially monocentric city region, dominated, also in terms of international recognition, by Birmingham, and it may use its leading role in the Eurocities local authority network to drive regional issues not least for its own advantage in international competition. There is a danger, of course, that the West Midlands region could become little more than a backyard for Birmingham's policies, a phenomenon most likely in monocentric regions. The case of Yorkshire and the Humber, by contrast, illustrates the importance of a sense of coherence and commonality also among essentially *competing* cities, if regionalization is to be effective and beneficial. A stronger and more visible political identity and capacity of the region could have provided greater scope for genuinely regional policies. This may include overcoming ideological-political divisions and historically grown localism, with divisions between the main cities and the formal region. Financial incentives are clearly a facilitator of such collaboration, at least for a limited, purpose-defined time, rather than a genuine commitment to, and acceptance of, the region. 'Marriages of convenience' will end after they have fulfilled their purpose, leaving no further commitment and need for mutual consideration. On the one hand it might well be beneficial that no fixed institutional arrangements have been cemented, so that existing purpose-built collaboration can be extended beyond its initial purpose and requirements. On the other hand, there is continuous uncertainty and jostling for position, which may be counterproductive as far as economic development is concerned.

Re-regionalization in Saxony-Anhalt: from state-installed to locally defined regions

Saxony-Anhalt is a structurally and geographically diverse *Land* in eastern Germany, with, not unlike Yorkshire and the Humber region in England, a largely rural, agriculturally dominated north and a more urbanized and industrialized southern half. The population of just under three million, about 20 per cent of the eastern German total population, is distributed accordingly. The two largest cities are in the south, one of which is the *Land* capital, Magdeburg, the other is the competing city of Halle. They share some 600,000 inhabitants, i.e. a fifth of the *Land* population. By contrast, the largest city in the northern part, Stendal (Map 7.2b), has just over 40,000 inhabitants (BBR, 2001, table 16) (Hoffmann *et al.*, 1991). The economic divisions within Saxony-Anhalt, which are reflected in the overall distribution of the main cities, however, suggest a somewhat more differentiated economic pattern than merely a crude north–south divide. Thus, seven economic entities (IWH, 1997) have been identified, which take into account local administrative and *Land* boundaries. The seven areas were eventually reduced to five to correspond with the five new planning regions by amalgamating the smallest two planning regions (Map 7.2).

The differing economic geography of Saxony-Anhalt became obvious in a cluster analysis of economic capacity in eastern Germany's five *Länder* (Barjak *et al.*, 2000). Thus, while it confirms the unique impact of the metropolis, Berlin, on the economically much weaker *Land* Brandenburg (see Chapter 6) as immediate, and eastern Germany as wider, hinterland, Saxony-Anhalt shows a quite mixed picture, and seems largely unaffected by the overarching effect of the major metropolitan centre. Its three largest cities, Halle, Magdeburg and Dessau together represent a mere 20 per cent of the 3.5 million population of Berlin, and they seem, therefore, less a dominant, but more an integral, part of the *Land* regional economic structure. The northern, rural part of Saxony-Anhalt shows distinctly common characteristics of peripherality and considerable structural economic deficiencies. This contrasts with the southern, more urbanized and successfully developing parts of the *Land*. The fact that these structural-economic divisions are represented in the new, bottom-up regional territorialization suggests at least some degree of appropriateness of the new boundedness vis-à-vis the economic geography. This should allow for common policy interests and concerns to emerge within each of the regions and thus the development of interlocal collaboration in the pursuit of common regional interests.

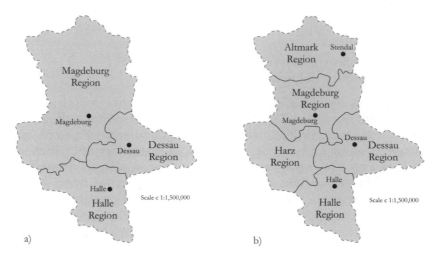

Map 7.2 (a) Old regions in Saxony-Anhalt until 1999; (b) New regions in Saxony-Anhalt since 1999.

Table 7.1 Economic and new planning regions in Saxony-Anhalt

Economic region (planning region)	Population	Structural feature	Main urban centre (population)
Altmark (*Altmark*)	253,000	rural, peripheral, stagnant	Stendal (46,000)
Anhalt (*Anhalt–Bitterfeld–Wittenberg*)	577,000	semirural with some suburbanization, (old) industrial	Dessau (93,000)
Halle, Merseburg, Saale-Unstrut (*Halle*)	497,000	densely populated, urbanized, strong economy (tertiary)	Halle (290,000)
Harz (hill country) and Mansfelder Land/Sangerhausen (*Harz*)	564,000	semirural, tourism	Halberstadt (42,000)
Magdeburg (*Magdeburg*)	635,000	semirural with some suburbanization, industrial-tertiary, *Land* capital function	Magdeburg (265,000)

Source: After IWH (1997).

The administrative structure of Saxony-Anhalt was essentially a quick fix, put in place in its present form between 1993 and 1995, with 1994 seeing a fundamental reorganization of the inherited small patterned administrative structure. Thus, the number of *Kreise* (upper tier local authorities) was almost halved from 37 to 21, and the smallest of the 1,300 or so local authorities were encouraged to pool some of their tasks to achieve greater efficiency in administration and service delivery. The target minimum number of inhabitants is 5,000 (MRLU S-A, *Land* Development Report (*Landesentwicklungsbericht*), 1996). By that time, the supra-local planning regions had also been established by simply giving this role to the three Land Government Offices situated in the three main cities of Magdeburg, Dessau and Halle. Their regional development programmes were adopted in 1995. The second *Land* elections in 1994 brought a shift in political control towards the left, and with it a questioning of the top-down approach in regionalization. Instead, a more bottom-up, locally based approach gained in favour and was applied to the regionalization of *Land* economic development and investment. This included the idea that municipalities would team up on the basis of common structural features and interests and thus define the new planning regions. Naturally, this new emphasis and regrouping affected the contents of the associated regional development plans. This process of bottom-up re-regionalization was completed in 1999 (see also MRU Saxony-Anhalt, 1999).

The initial arrangements for regionalization in Saxony-Anhalt were part of a fast track general institutionalization process across eastern Germany. In all five *Länder*, relevant legislation for defining regions and implementing regional planning and policy was passed between 1991 and 1993 (Müller, 1995). That is up to three years *after* unification, when many of the new economic structures had already taken clearer shape. Between four and five planning regions have been established in each *Land* by the respective *Land* governments, following established western German practices. In the absence of reliable and sufficiently detailed data, regions could often be defined merely on the basis of good estimates of, mainly urban focused, functional economic spaces. There was no time to wait for more reliable data to emerge over the years in the face of intense and immediate pressure from development processes. These were to be checked by introducing, rapidly, the proven western German system of spatial planning and policies, wholesale. Requiring orderly institutional response to development pressures, ready-made, western planning concepts and institutional practices were adopted almost without question. This included the use of urban centres and their functionally dependent hinterlands as the territorial basis of a 'region'. The limited reliability and availability of data, and having to rely on the indicators of the inherited socialist space economy, made such geography inherently questionable. The rapid structural changes and subsequent functional territorial realignments between cities and regions soon cast further doubts on the appropriateness of this imposed pattern of planning regions.

Table 7.2 Provisions for regional planning in the five new *Länder* in eastern Germany

Land	Territorial delimitation	Administration of planning regions	Initial legislation in
Mecklenburg-Westpomerania	• Functional areas of the four main cities, based on 'rough estimates'.	• Regional Planning Associations as municipal collaboration, offices for spatial planning and development run jointly by *Land* and local government.	1992
Brandenburg	• Functionally linked areas, five sector-shaped regions stretching from Berlin in centre to outer border.	• Regional Planning Associations as intermunicipal collaboration.	1992
Saxony-Anhalt	• *Land* Regional Offices areas, based on the three main cities as seats of Offices, • 5 new economy-based regions since.	• Regional representatives at *Land* Regional Offices, • Regional Planning Associations as intermunicipal collaboration since 1999.	1992
Saxony	• *Land* government established 5 planning regions on basis of urban-functional spaces, run by Planning Associations.	• Municipally run Planning Associations under *Land* guidance.	1992
Thuringia	• *Land* determined regions based on existing functional linkages, boundaries follow administrative areas.	• Regional Planning Associations, run in close collaboration with *Land*, *Kreise* and unitary urban authorities as main municipal players.	1991

Source: Land Planning Acts, Land Development Reports.

Table 7.2 summarizes some of the key provisions for regional policy making in the five *Länder* showing where Saxony-Anhalt fits in between strictly top-down approaches on the one hand, and the locally based bottom-up alternative on the other. The various *Land* Planning Acts provide the basis of all arrangements for regional planning. It becomes obvious that there are a number of similarities, due to the provisions made by federal law, but there are also distinct variations. All *Länder* have passed their regional planning legislation, which uniformly contains provisions for regions and related regional policies, with key objectives like 'facilitating indigenous growth' and establishing a 'functional region'. Differences exist in the nature of, and provisions for, the vertical relationships between central (*Land*) governments and the sub-*Land* regional administration. This involves, in particular, the role and standing of the regional planning associations as bodies specifically given the task to devise regional development plans (see also Chapter 5). Sitting between *Land* and local government, the respective balance in influence between *Land* governments, e.g. through the *Land* regional offices, and local government will circumscribe the region's nature as a top-down or bottom-up entity. In Saxony-Anhalt, for instance, at least initially, the emphasis was on top-down *Land* control of regional matters, through the *Land* regional offices, because this allowed the required quick fix implementation of the institutional-territorial structures (see Müller, 1995). After the discrepancies with the changing actual functional relationships became apparent, by the mid-1990s, a review of the existing, somewhat simplistic-generalist arrangement was undertaken. The territorial-administrative structures required greater differentiation and responsiveness to new economic reterritorialization, and a more indigenously defined, bottom-up approach to regionalization seemed more likely to produce the goods, using audits of regional economic potentials across the *Land*. Also, there was now more diverse and reliable data available. As a result, five new economy-derived regions emerged, two of which were merged to ensure a workable minimum size. Essentially, these regions are understood as 'spaces of cooperation with flexible boundaries' in response to changing policy requirements and/or priorities (Müller, 1999: 604). The result has been a re-regionalization following an increasing number of planning regions from three to four combined with de-scaling of institutional responsibility for regional planning from the centre-regional to the local–regional sphere. Thus, since 1998, local government, through the planning associations, is responsible for all regional planning (*Land* Planning Act 1998), instead of the *Land* government offices.

Nevertheless, all these devolved bodies operate within the *Land*-defined parameters, even the informal relationships. They are being channelled through new informal arrangements such as Regional Conferences and Regional Forums (see Figure 7.1), serving as platforms for informal collaboration between groups of local authorities with perceived common interests. Creating Regional Development Concepts or Regional Conferences is encouraged nationally by the federal government (BFLR, 1997) as a means of facilitating interlocal

Spatial Scale	'Old' *Land*-defined Regionalization	'New' Locally-defined Regionalization
Land	*Land* Planning Act (1992, 1996 amendments), requests *Land* Development Programme (strategic development aims)	*Land* government decides on devolving regionalization to the local level (1995) • *Land* guidance to local level on creating Regional Development Concepts (1995)
Region	'Old' Region simply defined as *Land* Government Office area (Regierrungsbezirk), i.e. from above Regional Development Programmes accepted as end result of 'old regionalization' (1996) Revision of Regional Planning Act (1999/2000)	'New' Region = voluntary association (collaboration) of local authorities (counties and unitary urban authorities) • Making of Regional Development Concepts and Regional Action Programmes as informal (statutorily non-binding) declarations of intent.
Local and regional actors in municipalities, regions, administrations, institutions, organizations, etc.		

Figure 7.1 Overview of the organization and operation of regional planning in Saxony-Anhalt.

Source: After information by Dessau City Council, Planning Department.

collaboration within functionally defined regions (Scholich, 1995), i.e. on the basis of affinity of interests among neighbouring local authorities. Such concepts are not part of the formal planning and policy-making, but they emerged as desirable additions to overcome some of the system's formal rigidities. In eastern Germany, the changes are leadind to more responsive policy making with regard to cultural, historic and geographic identities, and economic realities, than was feasible within the first round of top-down imposed formal regions. However, with a new territorial-administrative system established in eastern Germany in 1990, subsequent modifications to administration and responsibility had to retain existing administrative boundaries to avoid yet another reorganization vis-à-vis a public weary of yet more change (Schmitz, 1995). Reterritorialization was thus only possible through the combination of existing administrative local government units (both *Kommunen* and *Kreise*). The resulting picture of regionalization varies according to each *Land* government's willingness to allow devolutionary rescaling of development planning and policy which happened in Saxony-Anhalt in 1998 through a Statutory Act (Land Saxony-Anhalt, 1998).

The new arrangements for bottom-up regionalization are illustrated for the newly created region of Anhalt-Bitterfeld-Wittenberg – the names refer to the historic region of Anhalt, and two medium-sized cities – in the east of the *Land*, and this is also illustrated in Figure 7.2, see p. 197). This new region emerged from the previous Dessau region which took its name from that of its main city. Interestingly, the name of the revised region no longer refers to Dessau, which is a gesture towards the neighbouring rival city of Wittenberg (Dessau Planning Dept., July 2000). The diagram shows the prodedural changes to institutionalizing a planning region from a top-down to bottom-up approach after 1995. This shift was outlined in a guidance to the relevant local government bodies. The change in emphasis also reflects an essential shift in the understanding of the meaning of 'region'. Prior to 1995, 'regions' at the sub-*Land* level simply meant the three *Land* Government Office areas (*Regierungsbezirke*), and all (statutory) regional initiatives went through there. Since then, all such responsibilities have moved to the new style regions, i.e. the regional planning associations as the bodies responsible for representation. Thus, there is now a more immediate strategic input by those local authorities (upper tier and unitary urban authorities) which agreed to join forces in each of the new self-defined planning regions. The new-style regions are thus essentially locally decided, voluntary collaborative arrangements, but with clear statutory obligations, primarily designing and implementing Regional Development Concepts. The only compulsion for the higher tier local authorities, *Kreise* and unitary urban muncipalities, is to be part of such an association, although which one is left to their own choice.

With new boundaries emerging as a result of re-regionalization, ways need to be found for possible collaboration between neighbouring planning regions that happen to share a common economic area. Various possibilities are illustrated in Figure 7.1. The economic region with the three cities Halle, Leipzig and Dessau is of particular interest here, because it is separated by the *Land* boundary between Saxony and Saxony-Anhalt. Any effective regional development policy and planning would thus be expected to transcend this boundary. Indeed, the two states formally agreed in the late 1990s to facilitate cooperation in this area. It is noteworthy that such a formal act was deemed necessary and reflects the strongly formal nature of spatial planning. Collaborative work between the two sides is not made easier by the different *Land* constitutional arrangements and planning-political procedures and objectives which impact upon the regionalization processes. The result is an increased density and complexity of relationships between a wide range of local, regional and *Land* actors, leading to thick layers of differently scaled and bounded regional territories, each with their own responsibilities and associated aims. As Benz and Fürst (1998) point out, in this case, institutionalization is particularly difficult for that reason, leaving informal arrangements (which, of course, can easily be withdrawn or amended) the most feasible option.

The three planning regions covering this economic area are represented by the three steering committees of their respective Regional Forums on the joint Regional Conference covering the much wider inter-*Land* economic region of Halle-Leipzig-Dessau. These steering committees of the Regional Forums also participate in the *Land*-wide umbrella Forum 'Central Germany', which brings together academic and practical-administrative expertise and experience. The forums lead on to much smaller-scale, project-centred development strategies, such as 'regional innovation strategies'. This focus on projects is typical of informal collaboration, because it provides the necessary common purpose. Each such strategy addresses a single-issue policy project with its, generally, quite narrowly defined territorial boundedness.

Not surprisingly, therefore, there is no reference to planning or development narratives. The emphasis is distinctly on operationalizing pragmatic policy goals.

The system of bottom-up regionalization therefore depends heavily on the informal meetings of local council delegates on the various 'forums' and 'conferences'. Although the framework for these meetings has been formally instituted, there is much less control over the nature, conduct and competence of delegates of these negotiations or, indeed, their transparency and democratic scrutiny. Negotiations take place between the various regional actors, whether between or within regions, with both including horizontal and vertical relationships between actors and administrative hierarchies respectively. It is thus not surprising that these processes are seen as an important challenge to the coordination and operationalization of regional policy objectives (see Benz, 1996, 1998; Benz and Fürst, 1998; Fürst and Schubert, 1998). This includes in particular the conduct and rationale of negotiations between the formal, statutorily backed avenue of regional planning as a formal part of the planning hierarchy, and the optional, much less formally established regional development policies (Regional Development Concepts, Regional Action Programmes). Much of the implementation of these planning policies, however, ultimately depends on the availability and targeting of capital investment. As most of the capital originates from the *Land* government (with federal support), a close coordination between devolved regional development objectives and projects, and *Land*-controlled investment programmes, is essential. However, there still seems some way to go to link the two effectively.

The statutory basis of new regionalization is the 1998 Land Planning Act which establishes the dual system of regionalization: formally institutionalized and informally arranged regions (see Figure 7.1). The former encompasses the tier of formal development and land-use plans and specifies the organization and respective responsibilities of the *Land* and the five Regional Planning Associations responsible for regional planning.

The likely spatial incongruities in interests may reduce the efficacy of policies and thus the achievements of the whole system of regionalization. In addition, there are differing arrangements for intraregional cooperation (mechanisms) between actors and subsequent policy formulation. Thus, while all planning regions in Saxony-Anhalt use Regional Conferences as round table negotiation platforms, the degree to which these are institutionally formalized varies (Figure 7.2). Similarly, their visibility differs within and without the region. Having a registered office as a physical point of

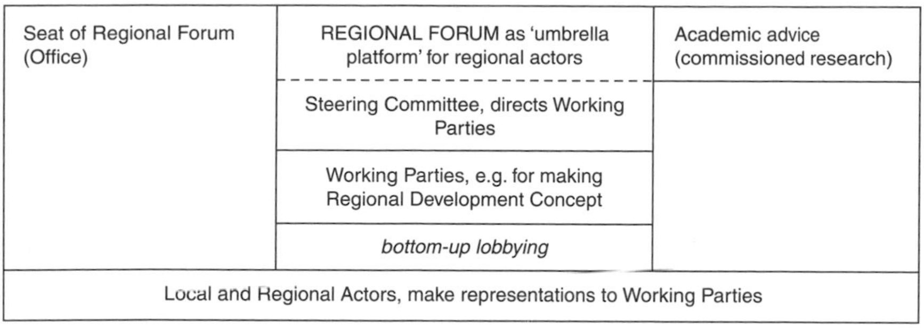

Seat of Regional Forum (Office)	REGIONAL FORUM as 'umbrella platform' for regional actors	Academic advice (commissioned research)
	Steering Committee, directs Working Parties	
	Working Parties, e.g. for making Regional Development Concept	
	bottom-up lobbying	
Local and Regional Actors, make representations to Working Parties		

Figure 7.2 Organizational structure of the regional forum Anhalt–Wittenberg–Bitterfeld.

Source: After information by Dessau City Council, Dept. of City Development.

access, for instance, provides an institutional presence and visibility in the public space, which goes beyond merely an existence as an invisible part of public administration. The legal status of these Conferences also varies from a registered non-profit private organization (Harz Region) to an association of local governments with distinct government-institutional elements. Thus, in the latter case, the planning association takes over certain agreed local tasks (e.g. local development planning and policy). In any case, all regions use external consultants and planning bureaux for a more efficient design of the Regional Development Concepts.

However, there are two sides to the coin, informality and the underlying absence of institutional responsibilities and legitimation (and the associated political pressures). For instance, by their nature, such arrangements exercise less pressure to produce results and be a firm part of a joint enterprise. Everyone can 'jump ship' or demand changes to policy goals or policies as such at any time. As a result, such formally non-committal arrangements are inherently volatile and hesitant, because they usually need to pursue the lowest common denominator to keep everyone on board. This situation may be a particular challenge to polycentric regions which lack a dominant and thus naturally leading larger urban authority which drives the agenda. In the cases where several equally sized local authorities compete for influence and pursuit of *their* interests, as in the cases of the Saxony-Anhalt and Yorkshire and the Humber regions, the willingness to compromise may be much lower. The absence of a high profile, registered office in some of the regions may be seen as a sign of the relevant local authorities' reluctance to grant the region too much authority and institutional standing, and establish too much of a sense of permanence, while weakening their own scope to manoeuvre in regional policy.

These difficulties have been raised in a study by Benz and Fürst (1998) on regionalization in Saxony-Anhalt. They highlight the challenges placed on local policy makers in particular, to put into operation a system of regional planning and policy making, which is less formalized and institutionalized, and thus places the onus of responsibility clearly on the participating local authorities. Among the advantages, the authors point to the positive impact of informal regional arrangements on the quality of communication between municipalities within and across government hierarchies. The difficulties they identified mirror those which emerged from the case studies examined here, in particular, the political pressures on local government to produce instant policy results as political capital. Then there are anxieties about losing local autonomy, difficulty in involving the business community, concerns regarding the financial responsibilities for regionalization tasks, and the general unclear legitimacy of informal regionalism (Benz and Fürst, 1998: 2).

Nevertheless, despite their obvious individuality and difference, the five new planning regions demonstrate important commonalities, such as clearly laid down avenues and opportunities for informal negotiations between various regional actors. So far, however, regional conferences have proven less effective in establishing a broader base of interest within the regions and informing policy makers (in the steering committees). Much is still driven by the government sphere and established civil service practices, rather than the wider community. Another common factor, is the dominant role of the *Kreise* with no independent, individual representation of smaller and rural local authorities and other actors. Generally, the emphasis is still clearly on 'government' rather than 'governance'. Differences between the five regions in Saxony-Anhalt include the degree of external visibility for regions as official entities, e.g. through regional offices as fully fledged business addresses and contact points. This is tied in with differing organizational arrangements ranging from forming a voluntary organization and thus a distinct institutional personality, to remaining a loose cooperation of local authorities without a distinct hands-on presence. This varying degree of formalization and institutionalization is the main difference between the regions' methods of operation. Benz and Fürst (1998) identify the absence of an HQ-style office as a major deficiency. The lack of perceived realness and accessibility for the various actors may be part of this. It seems that the regions are still in search of a suitable form of organization between traditional fixed institutionalization and more openness and flexibility to allow for more bottom-up decision making, albeit at the expense of coherence and procedural ease. Much of the policy making and planning processes is dealt with by steering groups, suggesting a distinctly technocratic approach with less transparency than might be desirable in the interest of democratic legitimacy. 'Insufficient institutionalization is in many cases considered a shortcoming' (Benz and Fürst, 1998: 3). This attitude may well reflect the inherited strong tradition in institutionalization and hierarchical administrative organization of decision making, particularly in eastern Germany, because *inter alia* it absolves bureaucrats of responsibility, a concern wide-spread under the former socialist system. The lack of experience may induce insecurity and uncertainty among the newly *formally* empowered regional/local actors. As a result, a more interventionist, hands-on role of the *Land* government may be seen opportune for the conduct of an effective form of regionalization which should also include non-governmental actors (Fürst, 1994) as a bolder step towards governance.

So far, a certain lack in democratic legitimacy may be identified. The difficulty is to connect conventional institutionalized forms with the new informal arrangements, because the former enjoy greater credibility and power than the latter. This mirrors the long-established competition between 'official' formal planning and the non-statutory forms of policy making, e.g. economic development policies (Herrschel, 1997, 1998). The employment of planning bureaux and consultancies to tackle development planning adds further to the sense of technocracy rather than democratically based governance. The fact that 'regional' projects often resemble little more than a basket full of the participating local authorities' pet projects, so as to please everyone involved, further adds to the feeling of window dressing rather than genuine strategic decisions and policies for the *region*. The outcome is thus an amalgamation of individual, locally managed *local* projects, simply labelled as 'regional', but not based on a genuine regional strategy and decision-making process. In order to lend regional policies some teeth, it is important that state investment policy and support to local authorities underpin the agreed regional strategies to reward participating actors. This, as Benz and Fürst (1998) stress, would include improved interdepartmental cooperation within the *Land* government. Currently, the provision of capital grants to business through joint federal-*Land* programmes and singular *Land* initiatives are administered by the government finance departments, and not regional bodies. In England, similar difficulties have been addressed quite successfully with the Regional Office approach. In addition, a need for establishing a learning process in managing informal regionalization is considered important to allow best practices to emerge, and this requires visibility and transparency of policy implementation.

New regionalization in Saxony-Anhalt – summary

The success of more locally determined regionalization, as attempted in the Saxony-Anhalt region, seems to depend considerably on solving some key problems by defining a clear strategy and putting it into practice. There are some inherent weaknesses in the policy-making system, in particular the absence of any financial muscle to focus local minds and facilitate the implementation of policy strategies. The problem, as stated repeatedly during interviews with key local and regional actors, lies essentially in the divided responsibilities between government departments, in particular, finance and economics. Policy incentives thus often remain uncoordinated. There is, therefore, an inherent centralized control element in regional planning and policy. Inevitably, such division increases friction and makes defining and implementing regionally defined policies more difficult. Local dependency on such regional grants causes resentment and friction, and makes non-cooperation, even more likely (Karrenberg, 1991). Trends towards devolving regional policies to the local level are therefore effectively counteracted by the centre's monopoly in effective instruments of regional policy making. This is particularly the case with the many smaller local authorities, often with fewer than 500 inhabitants and a part-time mayor. The resulting atomization of policy goals makes collaboration and compromise more difficult. Nevertheless, there was a general consensus among policy makers that further administrative reform towards larger local government units was unlikely, because people seem to increasingly resist losing their local identities. 'Too much has been lost already' was one comment describing the mood of the population (interview *Land* govt., June 1998). There are two main obstacles to a more rapid and visible development of regions: first, the lack of clear provisions for regionalism within the governmental hierarchy and, second, a relatively weak local input into regionalization by the many small *Kommunen* outside the main urban areas. The result, is often, different political and policy-making agendas by urban and (small) rural local governments.

Formal and informal planning and policy in polycentric regions – some concluding comments

The two case studies of regionalization processes in polycentric regions in England and Germany have illustrated the importance of localities and local interests within the internal organization and administrative-governmental operation of regions. This refers, in particular, to the scope for establishing homogeneous and genuinely region-based, effective policies. Interlocal competition and recognition, and acceptance, of a common regional agenda and its potential impact on local interests, vary considerably. Institutionalization per se does not lead automatically to a better integration between local interests for a regional purpose. Polycentric regions appear to require greater direction from above in the absence of an intraregional leadership by a dominant city, in order to drive a regional agenda and achieve compromises between, and coordination of, different subregional spheres and interests. Monocentric regions, by contrast, seem more likely to become self-directed, if largely under the auspices of the dominant city at the expense of the smaller municipalities, which leads to tensions. Locally based regionalization thus seems more feasible, albeit at the expense of more plural local input and hence democratic legitimacy of regional decisions as far as the underrepresented local areas in the region are concerned. Checks and balances, therefore, seem crucial, such as required consultation and negotiation, and the central state may have some (limited) role here. Polycentric regions seem here to promise more debate, but also potentially less decisiveness and boldness in policy making, when having to be content with the lowest common denominator in policy goals among the main political players of the region.

8 City-regional government and governance

In Chapters 5, 6 and 7 we examined in detail the institutional and policy challenges facing the regional scale in England and Germany. The analysis deals with the different structural conditions – external and internal – of monocentric and polycentric regions. *External* factors have been viewed as primarily the national constitutional framework and the role attributed to regions. This includes, in particular, the difference in devolution to the subnational level between unitary and federal states. Obviously, other factors, like historic-cultural and administrative legacies, will also play a role. *Internal* factors have been taken as local–regional relationships, i.e. the number and size of (urban) municipalities. In particular, the chapters focused on the difference between one single agglomeration controlling the region (monocentric), and several similarly large cities competing for influence (polycentric). The cases tackle issues of both the city region and wider regional scale. In Chapters 6 and 7 we saw how in the two German regions – one monocentric, one polycentric – the institutions set up at the time of unification in 1990 struggle to cope with contemporary economic challenges and considerable pressure for quick action in the first year or so. This has created administrative-territorial structures which do not always prove ideal. The story of the English regions is dominated by a national government that seems to know that the city and regional scale is important, but has not yet found a successful mix of devolved governance and central control.

The stories are different in the two internal structural contexts. In the monocentric, metropolitan regions there are similar issues of the multiplicity in institutions managing the dynamic core urban economy and its sub-regional and wider regional impacts. Both London and Berlin exemplify the need for effective city-regional governance, but in neither case could we confirm that the institutional mix is right, or policy set, effective. In the two other polycentric regions, there are common issues of relationships between cities and the regional scale and the need for different institutional and policy approaches in different subregions. All four cases certainly support an argument that European regional governance is likely to develop further in multilevel and asymmetrical directions.

As well as these similarities there are fundamental differences. One of the main issues facing regional planning in Berlin is the difficulty of cooperation between small units of local government. Regionalism here faces a problem of collective action; how to encourage cooperation around wider regional objectives and how to provide the incentives and regional concepts to facilitate such cooperation in the face of distinctly localist tendencies. Regional cooperation needs to be built from the bottom up, but this presumes the recognition of the local benefits of taking a regional view among local policy makers, particularly when it comes to local revenue. For the English regions, the perspective is very different. Both regional scale agencies and local governments are concerned primarily with conforming to central government policy guidance and fulfilling nationally agreed targets. While subregional and intraregional cooperation present problems both in the London region and in Yorkshire and the Humber, this has to be located in a top-down process of regionalization orchestrated by the centralized British state.

The constitutional and political context of regionalization has profound impacts. But this is not to say that we can deduce a respective natural form of regional governance from federal or unitary state contexts. Indeed, in Chapter 4 we saw some of the variety of city-regional and regional issues within these two types of European state. To understand differences between regions we need to look beyond basic constitutional difference. The German regions are relatively new regions. The post-war western German federal model was transferred to the new *Länder* in the east in 1990, but in very different economic and cultural conditions from those of the former West Germany. Economically weak regions understandably looked to counteract weakness and attract investment, and competitive local economies, represented by newly autonomous, powerful local authorities, have made intraregional cooperation difficult. The Berlin/Brandenburg region is a case in point. But it was not only the economic situation of the new *Länder* which shaped city-regional and regional plans. The democratization of subnational government also had profound impacts. In the early 1990s, newly autonomous local governments had little experience in managing private market forces, but political leaders and voters valued their new autonomy and were unwilling to submit to higher levels of decision making. There were strong historical and economic forces interacting with German federalism in our case studies, leading to distinct, complex conditions for the regions.

The impact of English centralism also mediated by other factors. The Yorkshire and the Humber case reveals a range of regional forms. Strong intraregional differences draw on historical, cultural and political identities, and are found behind, for example, the view of Leeds as a region-independent city state. Economic differences push regionalism in particular directions in stronger and weaker cities in the region. While Greater London had a history of metropolitan government, the South East and Eastern regions are new, artificial creations. Subregional differences, for example the relatively economically weak Thames Gateway, bring forward very different regional governance solutions.

Local economic forces, cultural and historical differences interact with formal constitutional differences in shaping regionalization. The interaction of these forces defines the scope for the emergence of the new networks and public–private sector connections that are identified by those who argue that a shift from government to governance is under way. How far do these cases support such an argument for governance? There is no doubt about the complexity of intergovernmental and public–private relationships. In the Berlin suburbs we saw new forms of horizontal cooperation between local governments and other actors. However, these emergent forms of sub-regional cooperation depend on the support of higher levels, in particular of the city of Berlin. The proliferation of partnerships in the English cases also gives support to the case for governance. The new government of London is radically different from previous metropolitan administrations. It has few staff, has engaged large numbers of external organizations in consultative networks and manages a development agency that incorporates business and other interests. The new government is thus inherently less bureaucratic and much more in tune with the idea of governance. However, hierarchical control is a defining feature of English regionalism. Central government controls budgets and policy, and creates and destroys metro-politan and regional government. Control of European funding also rests with the centre, i.e. the Department of Trade and Industry, and with Government Offices in the regions, not howeever with the regional bodies, such as the RDAs or local governments. The new governance in London exists alongside traditional ways of running the city region.

In Saxony-Anhalt we examined the shift from *Land*-defined regional policy and programmes controlled from above, to voluntary associations of local authorities in regions and new informal strategies. This can be seen as a shift from hierarchical *government* to horizontal *governance*. But the local, hori-zontal linkages are incomplete and the forms of regional planning lack formal status and, thus, the ability to steer resources from all levels of government. Smaller and more rural local governments are much less able to respond to the new style of regionalism. In our case studies many of the indicators of new governance are visible, but at best we would have to say that such processes are incomplete. In the English cases the tradition of central control is so deeply entrenched that further development towards govern-ance is difficult to foresee. In the Thames Gateway, central government drew back from developing local partnerships and imposed ministerial control over the subregion.

In both Germany and England, the overlapping new and older forms of regionalization create complex local and regional patterns of governance. One important issue in both countries is the continuing legitimacy of city and regional governance. Those subregions, which act as containers of European policy, have little legitimacy. European funding encouraged partnerships but there are doubts about the longevity of such forms of collaboration (ARL, 1998). Such temporary arrangements may not lay the foundations for longer-term regional identity. The new scale of regional-ization in Saxony-Anhalt is weakened by the lack of financial muscle, interlocal competitiveness, and the tendency of regional strategies to consist of little more than baskets of local projects. The technocratic style of regional planning is also a potential weakness. In Berlin/Brandenburg there was strong identity within local governments. Subregional plans had little support. The joint regional plan contained little more than a generalizing synopsis of local commitments. However, the concept of regional parks – a series of mixed use zones in the areas of development pressure around the city – does appear to be making some headway in obtaining recognition, as these regionally defined zones are acknowledged in estate agents' details, and local government opposition to the concept becomes less vociferous.

In London, the GLA's consultative processes may ensure some support for the London Plan, but such city-scale legitimacy can be undermined by central direction of regional policy. The success of metropolitan government will depend more on a successful resolution to arguments about transport investment. The legitimacy forged over the years by the consensus politics in SERPLAN and LPAC has been lost. Regional planning for the present continues as a process of dispute between local governments and the centre.

In some cases, cities and regions desire autonomy to build independent identities and legitimacy. Economically weak regions may well support much higher level intervention, if that brings funding. Chapter 3 examined the different scales of economic difference between regions. These differences clearly act on institutional and policy choices and there is no reason to think that inter- and intraregional disparities will lessen. Equally certain is a falling level of EU funding for German and English regions. The current programmes favour the development of capacities and economic potentials to allow regions to survive without EU funds. But, as we have seen, the development of cross-sector partnership and the shift from top-down government to governance and self-help is uneven. Differences between and within regions are continuing concerns. The Mayor of London wants the metropolitan area to retain more of the taxes it generates and to reduce the extent to which London's resources are redistributed to other English (and other British) regions. The mayor argues that the resources could (and should) be targeted at areas of deprivation within the city region, thus pointing to the intraregional structural heterogeneity and diverse policy needs. Similar arguments about tax and revenue distribution were heard in Berlin. The resolution of such fundamental issues for regional development depends on the intervention of higher levels of government. Equity across and between regions may not be able to be left to regional governments alone, not least because of the amount of resources required. In both England and Germany, regional and local initiatives need support from higher levels. Informal cooperation in the German regions needs financial

and other support from the *Land* governments. In England, central government will continue to manage the funding of regions at whatever scale.
Regionalization, in our case studies, presents collective action problems for
local governments but also continues to challenge higher level governments
to seek efficient and effective management at regional scales.

New regionalism?

The multiple perspectives on regionalization that we examined in Chapter
2 raised profound questions about the relationships between economies,
territory and governance. At the end of the 1990s many of these issues were
focused on debate about the 'new regionalism'. How far do the results from
these case studies support the new regionalist arguments?

The new regionalists see transformative potential in regional cooperation
and regional scale planning. In particular, they argue that regionalism can
bring environmental and equity benefits. Our English regions follow a rigid
national agenda on sustainable development. Thus, while environmental
issues are taken seriously at regional scale, this represents a national, rather
than a local agenda. The new RDAs are also responsible for managing central
government's regeneration policy which assists deprived areas and groups.
In the future, the government's SRB programme will end and the budget
will be transferred to the RDAs. The RDAs may well shift this money into
marketing and development projects to assist competitiveness. The current
emphasis on equity may well be weakened by this devolution of responsibility. In London and elsewhere, however, the dominant theme of regional
policy is competitiveness. It is economic, rather than equity or environmental
issues that drive the English regional agenda.

In Berlin/Brandenburg economic competition between the *Land* governments and between local governments dominates other policy objectives.
Economic competition therefore seems to be the driving force of regionalism in both countries. Competitiveness is the theme of European level
policy (as we saw in the ESDP). National policies for cities and regions, and
economic competition, dominate at regional and local levels. European
policy responds to global competition and Europe's cities and regions vie
with each other for advantage. Both cities and regions, however, need
support from higher levels if they are to tackle environmental and equity
issues successfully. The shift in EC programmes from regional to urban scale
targets intraregional disadvantage and provides a step in the direction of
locally (city) centred regionalism. The English SRB programme takes a
similar local area focus. While regional bodies are focused on competitive
advantage these higher level programmes tackle a broader agenda which
reflects a shift from conventional, general, welfare oriented regional policy
to more specific, competition focused measures of a 'workfare state'.

But the ambitions of the new regionalism are built on the mobilization
of new regional actors around integrative regional agendas. What is clear

from our cases is that competitiveness and environmental and equity planning at regional scale has to involve multiple levels of government and public and private networks. It is also clear that not all regions will follow the same direction. Some, particularly the more successful metropolitan regions, can develop the cross-sector networks to deliver broad policies, others will look to support from higher levels. What seems to work at regional scale are programmes that are incentivized (by European or national programmes or by perceptions of common advantage) or nationally directed. In all cases, a clear benefit from participation in this for the individual locality needs to be visible. Some commentators claim that traditional top-down regional planning is giving way to new, flexible forms of horizontal cooperation bringing forward associational forms of governance (see Allen *et al.*, 1998). Such claims underestimate the continuing roles of national and supranational government. There is certainly no one clear direction of change, or one privileged scale in the governance of cities and regions. Arguments about path dependence clearly form part of the explanation. English central–local government relations, the historic distrust of subnational actors, clearly shape current institutions and policies. An important question is how far devolution to Scotland and Wales develops and the extent to which lessons from those experiments impact on English regionalization. The case for city regions, supported by cities in Yorkshire and the Humber, will depend in part on the perceived success of London government. English regional devolution is constrained by past experience but the limited devolution introduces new institutions that may well contribute to further changes. The constraints of regional boundaries set down in 1990, clearly limit regional cooperation in Berlin. It may be that by the time of a new election in 2009 the benefits of regional cooperation will have become clear enough to override historic suspicion between Berliners and Brandenburgers. The success of regional planning initiatives could facilitate a break out from the impasse of the post-unification years.

European comparisons and lesson drawing

Comparative studies can offer potential insights into how regional identity and support for a city-regional scale of intervention may be built through particular processes and mechanisms of cooperation, planning and management. The case for the new regionalism would be all the stronger if we could point to progressive impacts of management at city region and regional scale. In our studies we encountered numerous officials and political leaders committed to better regional cooperation and more effective intervention at this scale. These individual efforts may well help steer the path of future regional development. While influential actors need to be taken into our understanding of city and regional development, at present there is only limited evidence of progressive policy taking effect at these scales.

Two types of lesson can, however, be identified from the English and German cases. First, there are institutional lessons about forms of cooperation. The emerging cross-border cooperation in Berlin may suggest potential institutional solutions for the governance of London's cross-border economic wedges. The voluntary association behind the Regional Development Concept may have lessons for English regions. Second, there are potential policy lessons and lessons about forms of effective regional policy. Regionalism needs strong images on which to build identity and progressive policy. One lesson from the case studies is that such processes take time. The Berlin regional park concept was opposed by local governments but, after five or six years opposition, resistance is becoming less intense. The Regional Development Concept can also build support for cooperation. These bottom-up initiatives contrast with the top-down direction of plans and planning in England and reliance on marketing images to create regional identities.

We examined in Chapter 2 the set of arguments about the scalar restructuring of economies and governance. The key element of this fundamental process of restructuring was argued to be the entrepreneurial urban region. Our review of European examples and the English and German case studies suggest that competitive city regions are just one of the forms of city and regional governance. Not all cities can aspire to this role and the range of responses to economic competition and environmental and social challenges in Europe undermines the insistence in the scale debate on the functional necessity of new urban governance. Economic forces interact with national constitutional factors and local politics. They also interact with historical and cultural contexts. Our results show the impacts of formal regionalization and the ability of actors to forge informal alliances. A limitation of comparative studies is that conclusions often point to national contextual differences which are unsurprising. However, comparing policy and planning developments at city and region scale focuses less on the formal systems of government and more on arenas of formal and informal policy making and on the mobilization of ideas around regional projects, and the role of individual actors or actor groups. The particular regions and initiatives we have analysed show how far the city and regional scale is contested between governmental and other actors, and that this applies to both types of state organization – centralized and devolved. For the present, the driving force behind regionalism is economic competitiveness, but this is expressed in different ways in differing institutional contexts and against varying cultural-political legacies.

Glossary

England

DETR/DTLR Department of the Environment, Transport and the Regions (1997–2001), Department of Transport, Local Government and the Regions (2001–).The government department responsible for regional and local government.

DTI The national government Department for Trade and Industry.

GLA Greater London Authority, established in 1999 as London-wide administrative body, under the leadership of the Mayor of London.

GLC The Greater London Council (1965–1986) responsible for strategic planning across the 33 London boroughs.

GOR Government Offices in the Region established in 1994. These include GOL (the Government Office for London), GO-East (for the eastern region), and GOY&H (for Yorkshire and the Humber).

LPAC The London Planning Advisory Committee (1986–2000) drew the London boroughs together to advise central government on strategic planning issues in the form of Regional Planning Guidance (RPG). RPG is a national system of plan making at regional scale. Strategic plan making in London has now passed to the Mayor of London and the Spatial Development Strategy (SDS).

RDA Regional Development Agencies established in England in 1999. These include SEEDA. The LDA (London Development Agency was set up in 2000) and, exceptionally, its board is appointed by the Mayor of London rather than central government.

SERPLAN The South East Regional Planning Conference that, until 2001, brought all subnational governments in the region into the process of regional planning. (See also LPAC.)

Europe

The term Europe is somewhat ambiguous in its general usage. Here, it is understood in its geographic sense, i.e. including 'western' and 'eastern' Europe.

COR   Committee of the Regions established under the (Maastricht) Treaty of European Union 1993 as a consultative forum for subnational governments.

EC   The European Commission comprising functional Directorates (DGs), DGXVI having responsibility for regional policy and the administration of regional development funds (ERDF).

EU   The European Union of member states.

European Structural Funds   Comprises the ERDF and the European Social Fund (ESF).

NUTS   Nomenclature of Territorial Units for Statistics (the NUTS acronym reflects the French *Nomenclature des Unités Territoriales*) attempts to provide a common classification of information at subnational level across the EU. NUTS are organized in a scalar hierarchy, with growing NUTS number corresponding with decreasing scale: NUTS 0 corresponds to the territory of the nation state, and NUTS 4 to the local level. NUTS 1 refers to the first level of subnational regions.

Germany

Bezirke   The city state of Berlin is divided into *Bezirke* but these have limited powers, and are, thus, not fully fledged local government units. This function is held by the Berlin Senate in its dual role as *Land* and city government.

BROG   The *Bundesraumordnungsgesetz* – the national spatial planning laws.

GDR   German Democratic Republic – the former East Germany.

Gegenstromprinzip   The unique German process of 'counterflow' of bargaining and negotiation between tiers of government.

Gemeinsame Landesplanungsstelle   This is the joint Berlin/Brandenburg regional planning body.

Grundgesetz   The national Basic Law, setting the scope of local autonomy.

Kommunen   Basic units of local government. The *Gemeinderat* is the local council and basic decision-making forum and local legislator. *Kommunen* can group into *Kreise* (a county scale) and Ämter (to cooperate on service provision on behalf both of local authority and as subregional agents of the *Land* government).

Land (*Länder*)   The 16 formal regions (Federal states) of Germany, each of which produces a strategic development plan the *Landesentwicklungsprogramm* as guidance for lower tier spatial planning.

Planungsregionen   Planning regions – subdivisions of the *Land* territory for the elaboration of formal plans, following the guidance of *Land* development plans.

Regierungsbezirk   The areas of local offices of a *Land* may also form units for regional planning, reflecting an essentially more top-down approach to regional planning.

Regionale Planungsgemeinschaften   Regional Planning Associations of local governments, whose main task is the production of a Regional Development Plan.

International perspectives

NAFTA   North American Free Trade Association (US, Canada and Mexico; ratified in the US in 1993) and which has spatial planning implications, in particular, in relation to transport planning.

OECD   Organisation for Economic Cooperation and Development, an international body promoting economic development and which involves subnational governments in producing good practice advice.

References

Abercrombie, P. (1945) *Greater London Plan*, HMSO.

Abu-Lughod, J. (1999) 'New York, Chicago, Los Angeles: America's global cities', Minneapolis: Minnesota University Press.

Ache, P. (2000) 'Cities in old industrial regions between local innovative milieu and urban governance – reflections on city region governance', *European Planning Studies*, 8(6): 693–709.

AEBR (Association of European Border Regions) (1996) *An Introduction to the Association of European Border Regions*, Enschede: mimeo.

AFLRA (Association of Finnish Local and Regional Authorities) (2001) Background on its structure and organization on its web pages under www.kuntaliito.fi/english/frmaineng.htm, accessed on December 2001.

Agnew, J. and Brusa, C. (1999) 'New rules for national identity? The Northern League and political identity in contemporary Northern Italy', *National Identities*, 1(2): 117–133.

Ahlke, B. (1997) 'Transnationale Zusammenarbeit in der Raumordnung: neue Ansätze, Programme, Initiativen', *Informationen zur Raumentwicklung* (6): 361–378, Bonn: BBR.

Aigner, B. and Miosga, M. (1994) 'Stadtregionale Kooperationsstrategien. Neue Herausforderungen und Initiativen deutscher Großstadtregionen', *Münchner Geographische Hefte*, vol. 71, Regensburg: Michael Lassleben.

Albers, G. (1998) 'Kompakte und durchmischte Städte in polyzentrischen Stadtregionen – Eine Einführung', in BMBau/empirica (eds) *Die Zukunft der Stadtregionen*. Documentation of a conference of this title, Hanover, 22–23 October 1997, pp. 51–55.

Albrechts, L. (1998) 'The Flemish diamond: precious gem and virgin area', *European Planning Studies*, 6(4): 411–424.

Allen, J., Massey, D. and Cochrane, A. (1998) *Rethinking the Region*, Routledge: London.

Altmarkkreis and Landkreis Stendal (eds) (2000) 'Regionales Aktionsprogramm Altmark-RAP II'. Salzwedel/Stendal: unpublished.

—— (ed.) (1994) *Post-Fordism: A Reader*, Blackwell: Oxford.

—— (1999) 'An institutionalist perspective on regional economic development', *International Journal of Urban and Regional Research*, 23(3): 365–378.

—— and Graham, S. (1997) 'The ordinary city, transactions of the Institute of British Geographers', *New Series*, 22: 411–429.

—— and Robins, K. (1990) 'The re-emergence of regional economies? The mythical geography of flexible accumulation', *Environment and Planning A*, 8(1): 7–34.

—— and Thrift, N. (1994) 'Living in the global', in A. Amin and N. Thrift (eds) *Globalization, Institutions, and Regional Development in Europe*, Oxford: OUP, pp. 1–22.

Anadyke-Danes, M. *et al.* (2001, for Regional Studies Association) *Labour's New Regional Policy: An Assessment*, Seaford: RSA.

Anderson, J. (1996) 'The shifting stage of politics: new medieval and postmodern territorialities?', *Environment and Planning D*, 14(2): 133–153.

—— and Hamilton, D. (1999) 'The separation/integration of "Economics" and "Politics" in border regions and cross-border development', paper presented to the RSA Conference *Regional Potentials in an Integrating Europe*, Bilbao, September 1999.

Antikainen, J. (2000) 'Challenge of operationalising the concept of urban networking as the development principle – the case of Finland', paper presented to 3rd EURS conference, Voss, September 2000.

ARL (Akademie für Raumforschung und Landesplanung) (1998) *Interkommunale und -regionale Kooperation. Variablen ihrer Funktionsfähigkeit*, Hanover: ARL.

ARP (Arbeitsgemeinschaft Regionale Planungsgemeinschaft) 'Prignitz Oberhavel (1997) Regionale Kooperation im Spannungsfeld zwischen Metropolis und äußerem Entwicklungsraum', Nachhaltige Entwicklung im Raum Berlin-Brandenburg: unpublished.

Aschauer, W. (2000) 'Zwischen Analyse und Politik. Zum Anwendungsbezug regionalwisseschaftlicher Forschung', *Informationen zur Raumentwicklung* (9/10): 589–598, Bonn: BBR.

Atkinson, R. and Moon, G. (1994) *Urban Policy in Britain*, Basingstoke: Macmillan.

Bache, I. (1999) 'Explaining variations in regional empowerment through EU structural policy: the case of the United Kingdom', paper presented to the Regional Studies Association Conference, Bilbao, 18–21 September 1999.

—— (2000) 'Government within governance: network steering in Yorkshire and the Humber', in *Public Administration*, 78(3): 575–592.

Bachtler, J. and Downes, R. (1999) 'Regional policy in the transition countries: a comparative assessment', *European Planning Studies*, 7(6): 793–808.

—— (2000) 'The spatial coverage of regional policy in central and eastern Europe', *European Urban and Regional Studies*, 7: 159–174.

Bagnasco, A. and Oberti, M. (1998) 'Italy: "le trompe-l'oeil of regions"', in P. Le Galès and C. Lequesne (eds) *Regions in Europe*, London: Routledge.

Baldersheim, H. and Ståhlberg, K. (1999) *Nordic Region-Building in a European Perspective*, Aldershot: Ashgate.

Balme, R. (1998) 'The French region as a space for public policy', in P. le Galès and C. Lesquesne (eds) *Regions in Europe*, London: Routledge, pp. 181–198.

Barjak, F., Franz, P., Heimpold, G. and Rosenfeld, M. (2000) 'Regionalanalyse Ostdeutschland: die wirtschaftliche Situation der Länder, Kreise und kreisfreien Städte im Vergleich', Institut für Wirtschaftsforschung Halle (ed.) *Wirtschaft im Wandel*, vol. 6, no. 2, pp. 31–55.

Barnes, W. R. and Ledebur, L. C. (1997) *The New Regional Economies*, London: Sage.

Baumheier, R. (1997) 'Raumordnungskonferenz Bremen/Niedersachsen. Regionale Entwicklung im Wechselspiel gesamträumlicher und teilräumlicher Ansätze', *Informationen zur Raumentwicklung*, (3): 161–166.

BBR (1999a) 'Modellvorhaben "Städtenetze". Neue Konzeptionen der Interkommunalen Kooperation', *Werkstatt Praxis*, no. 3, Bonn: BBR.

—— (1999b) 'Regionen der Zukunft – Regionale Agenden für eine nachhaltige Raum- und Siedlungsentwicklung', *Wettbewerbszeitung*, no. 2, Bonn: BBR.

—— (2000) 'Urban development and urban policy in Germany. An overview', Series *Berichte* (reports) no. 6, Bonn: BBR.

—— (2001) 'Aktuelle Daten zur Entwicklung der Städte und Gemeinden', Series *Berichte* (reports), vol. 8, Bonn: BBR.

—— (INKAR) (1998) *Indikatoren und Karten zur Raumentwicklung*, collection of statistics, Bonn: BBR.

—— (INKAR) (2000) *Indikatoren und Karten zur Raumentwicklung*, collection of statistics, Bonn: BBR.

Beauregard, R. and Pierre, J. (2000) 'Disputing the global: a sceptical view of locality-based international initiatives', *Policy and Politics*, 28(4): 465–478.

Beaverstock, J., Smith, R. and Taylor, P. (1999) 'A roster of world cities', *Cities*, 16: 445–458.

Benneworth, P. (2000) 'An initial assessment of the eight final regional development strategies', in *Regions: The Newsletter of the Regional Studies Association*, no. 225: 9–21.

Bennington, J. and Harvey, J. (1994) 'Spheres or tiers; the significance of trans-national local authority networks', in P. Dunleavy and J. Stanyer (eds) *Contemporary Political Studies*, Belfast: Political Studies Association, pp. 943–961.

Benz, A. (1996) 'Regionalisierung in Sachsen-Anhalt. Begleitforschung zur Regionalisierung in Sachsen-Anhalt. Bestandsaufnahmen von Regionenbildung und Entwicklung von Regionalkonferenzen und Untersuching von Möglichkeiten einer besseren Institutionalisierung von Regionen', report for state government of Saxony-Anhalt, Halle: University of Halle, unpublished.

—— (1998) 'German regions in the European Union. From joint policy-making to multi-level governance', in P. Le Galès and C. Lequesne (eds) *Regions in Europe*, London: Routledge.

—— and Fürst, D. (1998) 'Bericht zur kommunalisierten Regionalisierung in Sachsen-Anhalt', report to the government of Saxony-Anhalt, Halle: University of Halle, unpublished.

—— and Koenig, K. (1995) *Der Aufbau einer Region: Planung und Verwaltung im Verdichtungsraum Berlin-Brandenburg*, Baden-Baden: Nomos.

Berg, L. van der, Braun, E. and Meer, J. van der (1997) 'The organising capacity of metropolitan regions', *Environment and Planning C: Government and Policy*, 15: 253–272.

BFLR (Bundesforschungsanstalt für Landeskunde und Raumordnung) (ed.) (1995) 'Laufende Raumbeobachtung Europa. Daten zur Struktur und Entwicklung der Regionen der Europäischen Union', *Materialien zur Raumentwicklung*, vol. 71.

Biskup, M. (1994) 'Aktuelle Aspekte der Regional- und Landesplanung in den Neuen Bundesländern', in J. Domhardt and C. Jacoby (eds) *Raum- und Umweltplanung im Wandel. Festschrift für Hans Kistenmacher*, Kaiserlautern: Selbstverlag Universität Kaiserslautern, pp. 179–191.

Blotevogel, H. H. (2000) 'Zur Konjunktur der Regionsdiskurse', *Informationen zur Raumentwicklung*, (9/10): 491–506, Bonn: BBR.

BMRBS (Bundesministerium für Raumordnung, Bauwesen und Städtebau) (ed.) (1996) 'Raumordnung in Deutschland', Bonn: BMRBS, unpublished.

Bomberg, E. and Peterson, J. (1998) 'European Union decision making: the role of sub-national authorities', *Political Studies*, 46(2): 219–235.

Bordlein, R. (2000) 'Die neue Institutionalisierung der Region: das Beispiel Rhein-Main', *Informationen zur Raumentwicklung*, (9/10): 537–548, Bonn: BBR.

Boyer, R. (1998) *The Regulation School: A Critical Introduction*, New York: Columbia University Press.

Bradford City Council (1999) Interview with Economic Development Unit, April 1999.

Brenner N. (2000) 'The urban question as a scale question: reflections on Henri Lefebvre, urban theory and the politics of scale', *International Journal of Urban and Regional Research*, 24(2): 361–378.

—— (2002) 'Decoding the newest "metropolitan regionalism" in the USA: a critical overview', *Cities*, 19(1): 3–21.

—— and Heeg, S. (1998) 'Leistungsfähige Länder, konkurrenzfähige Stadt-regionen? Standortpolitik, Stadtregionen und die Neugliederungsdebatte in den 90er Jahren', *Informationen zur Raumentwicklung*, (10): 661–672.

Brown, C. and Cäniels, M. (1997) 'Border regions in Europe: exploring the myth and mystery', paper presented to the RSA conference, Frankfurt/Oder, September 1997.

Bufalica, A. (1995) 'Die neuen Nachbarn, Verwaltungsbeziehungen zwischen Berlin und seinen Nachbarkommunen'. Series 'Beiträge aus dem Fachbereich I der Fachhochschule für Verwaltung und Rechtspflege' (FHVR), Berlin: FHVR.

Burns, M. C. and Cladera, J. R. (2000) 'Barcelona and its role as the capital of an increasingly polycentric regional urban system', paper presented to the European Urban Research Association, Spring Workshop, Dublin 13–15 April 2000.

Cabinet Office (2000) *Reaching Out; The Role of Central Government at Regional and Local Level*, London: PIU.

CEC (1990) *Commission of the European Community: Europe 2000*, Luxembourg.

—— (1991) *Europe 2000: Outlook for the Development of the Community's Territory*, Brussels: CEC.

—— (1992) *Urbanisation and the Functions of Cities in the European Community*, Brussels: CEC.

Cheshire, P. and Gordon, I. (eds) (1995) *Territorial Competition in an Integrating Europe*, Aldershot: Avebury.

—— (1996) 'Territorial competition and the logic of collective in(action)', *International Journal of Urban and Regional Research*, 20: 383–399.

Church, A. and Reid, P. (1998) 'Urban power, international networks and competition: the example of cross-border co-operation', *Urban Studies*, 33(1): 297–318.

—— (1999) 'Political space across the channel tunnel', *Regional Studies*, 33(7): 643–656.

City of Turin (*c.* 2000) 'Torino Internazionale – Strategic plan for the promotion of the city' (leaflet), Turin: unpublished

Clark, T. N. and Hoffman-Martintot, V. (eds) (1999) *The New Political Culture*, Boulder: Westview Press.

Clarke, S. and Gaile, G. (1997) 'Local politics in a global era: thinking locally, acting globally', *Annals of the American Academy of Political and Social Science*, 651: 28–43.

Cochrane, A. (1993) *Whatever Happened to Local Government?*, Buckingham: Open University Press.

—— and Jonas, A. (1999) 'Reimagining Berlin. World city, national capital or ordinary place?', *European Urban and Regional Studies*, 6(2): 145–164.

Cooke, P. and Morgan, K. (1994) 'Growth regions under duress: regional strategies in Baden-Württemberg and Emilia Romagna', in A. Amin and N. Thrift (eds) *Globalisation, Institutions and Regional Development in Europe*, Oxford: Oxford University Press.

—— (1998) 'The regional innovation system in Baden-Württemberg', *International Journal of Technology Management*, 9, 394–429.

Corporation of London (2000) 'London–New York Study final report', Economic Development Unit, London: Corporation of London.

Cox, K. R. (ed.) (1997) *Spaces of Globalization*, New York: Guilford Press, pp. 137–166.

Cross, C. (1981) *Principles of Local Government Law*, London: Sweet & Maxwell.

CSD (Committee On Spatial Development) (1999) *European Spatial Development Perspective*, Tampere: CSD.

Danielzyk, R. (1995) 'Regionalisierte Entwicklungsstrategien – "modisches" Phenomen oder neuer Politikansatz?', in A. Momm, R. Löckner, R. Danielzyk and A. Priebs (eds) *Regionalisierte Entwicklungsstrategien. = DVAG – Material zur Angewandten Geographie*, Vol. 30, Bonn: Irene Kuron.

—— (1998) 'Leigt die Zukunft in den Regionen?', paper presented to the Thüringische Regionalplanertagung (Thuringian Conference of Regional Planners), Erfurt, June 1998.

—— (1999) 'Regionale Kooperationsformen', *Informationen zur Raumentwicklung*, (9/10): 577–587.

—— and Wood, G. (2000) 'Innovative strategies of political regionalization. The case of North-Rhine Westphalia', paper presented to the 3rd EURS conference, Voss, Norway, September 2000.

Danson, M. W., Fairley, J., Lloyd, M. G. and Turok, I. (1999) 'The European Structural Fund Partnerships in Scotland: new forms of governance for regional developments?', *Scottish Affairs*, (27): 23–40.

Department of Economic Development (1999) *Strategy 2011*, Report by the Economic Strategy Review Steering Group.

DETR (Department of the Environment, Transport and the Regions) (1997) *Building Parnerships for Prosperity*, London: DETR.

—— (1998a) *Draft Statutory Guidance for RDAs*, London: DETR.

—— (1998b) *Building Partnerships in the English Regions*, London: DETR.

—— (2000) 'Regional planning guidance for the South East of England. Public Examination May–June 1999', report of the panel: unpublished.

Deutscher Städtetag (1995/6) 'Interkommunales Handeln in der Region', working paper, Cologne: unpublished.

Dieleman, F. and Faludi, A. (1998) 'Polynucleated metropolitan regions in northwest Europe: theme of the special issue', *European Planning Studies*, 6(4): 365–377.

DIW (Deutsches Institut für Wirtschaftsforschung) (ed.) (1995) *Transferleistungen in die neuen Bundesländer und deren Wirtschaftliche Konsequenzen*, Berlin: Duncker & Humblot.

—— (ed.) (1997) 'Gesamtwirtschaftliche und unternehmerische Anpassungsfortschritte in Ostdeutschland', *DIW Wochenbericht*, 64(23): 549–579.

—— (ed.) (1999) 'Gesamtwirtschaftliche und unternehmerische Anpassungsfortschritte in Ostdeutschland', *DIW Wochenbericht*, 66(23): 419–445.

DMEE (Danish Ministry of Environment and Energy) (2000) 'Planning report', Copenhagen: unpublished.

DOE (Department of the Environment) (1992) 'East Thames corridor: a study of development capacity and potential', unpublished.

—— (1994) *Regional Planning Guidance for the South East RPG9*.

—— (1995) *The Thames Gateway Planning Framework RPG9a*, also in HMSO internet summary, www.official-documents.co.uk.

DTLR (Department of Transport, Local Government and the Regions) (2000) 'Action to boost Thames Gateway', news release 2000/0369, 22 May.

Duchacek, I., Latouche, D. and Stevenson, D. (eds) (1998) *Perforated Sovereignties and International Relations: Transsovereign Contacts of Subnational Governments*, Westport, CT: Greenwood Press.

Dunford, M. (1994) 'Winners and losers: the new map of economic inequality in the European Union', *European Urban and Regional Studies*, 1(2): 95–114.

—— (1997) 'Regions and economic development', in P. Le Galès and C. Lequesne (eds) *Regions in Europe*, London: Routledge.

EC (European Commission) (1994) *Europe 2000+. Cooperation for European Territorial Development*. Luxembourg: Office for Official Publication of the European Communities.

—— (1997) *Fifth Periodic Report on the Regions: Summary of Main Findings*, Brussels: EC.

—— (1998) *Towards an Urban Agenda*, Brussels: EC.

—— (1999a) 'Sixth periodic report on the regions', *Inforegio*, Luxembourg: Office for Official Publication of the European Communities, accessed under www.inforegio.cec.ei.int/wbover/overcon.oco2a_en.htm, on 4 November 1999.

—— (1999b) www.inforegio.cec.ei.int/wbover/overcon.oco2a_en.htm, accessed 4 November 1999.

Elsner, W. (2000) 'Regionalisierung und Neuer Regionalismus. The big divide: Neoliberalismus oder Proaktive Regionalpolitik', *Informationen zur Raumentwicklung*, (9/10): 575–558, Bonn: BBR.

Eltages, M. (1997) 'Programmorientierte transnationale Zusammenarbeit als Aufgabe der europäischen Strukturpolitik', *Informationen zur Raumentwicklung*, 6, pp. 379–386.

Eurocities (1998) Eurocities For An Urban Policy (www.eurocities.org/euroact/action/html).

Faludi, A. (1997) 'European spatial development policy in "Maastricht II"?', *European Planning Studies*, 5: 535–543.

—— (2000) 'The European spatial development perspective – what next?', *European Planning Studies*, 8, pp. 237–250.

Florida, R. (1995) 'Toward the learning region', *Futures*, 27: 527–536.

Foster, K. (2001) 'Regionalism on purpose', Lincoln Institute Policy Focus Reports PF011, Washington: Lincoln Institute.

Friedmann, J. (1999) *World City Futures: The Role of Urban and Regional Policies*, mimeo.

Furst, D. (1994) 'Regionalkonferenzen zwischen offenen Netzwerken und fester Institutionalisierung', *Raumforschung und Raumordnung*, 52(3): 184–192.

—— and Schubert, H. (1998) 'Regionale Akteursnetzwerke. Zur Rolle von Netzwerken in regionalen Umstrukturierungsprozessen, *Raumforschung und Raumordnung*, 56(516): 352–362.

Gachelin, C. (1998) 'The ambition of Eurocities', *Urbanisme Hors Series*, 10: 18–25.

Gaffikin, F. and Morrisey, M. (2001) 'The other crisis. Restoring competitiveness to Northern Ireland', *Regional Economy*, 16(1): 26–37.

Georgiou, G. (1999) 'The dynamics of the community support framework for Greece and its contribution to regional convergence', *European Planning Studies*, 7: 665–676.

Gibbs, D., Jonas, A., Reimer, S. and Spooner, D. (1999) 'Not enough partners and too many plans: governance and institutional capacity in the Humber sub-region', paper presented to the RSA international conference, Bilbao, 18–21 September 2001).

—— (2001) 'Governance, institutional capacity and partnerships in local economic development: theoretical issues and empirical evidence from the Humber sub-region', *Transactions*, 26: 103–120.

Giordano, B. (2001) 'The contrasting geographies of "Padania": the case of the Lega Nord in northern Italy', *Area*, 33(1): 27–37.

GLA (2001) Towards the London Plan, Greater London Authority http://www.london.gov.uk/approot/mayor/strategies/sds/towardslonplan.jsp.

Gleisenstein, J., Klug, S. and Neumann, A. (1997) 'Städtenetze als neues "Instrument" der Regionalentwicklung?', *Raumforschung und Raumordnung*, 1: 38–47.

Gödecke-Stellmann, J., Müller, A. and Strade, A. (2000) 'Konkurrenz und Kooperation. Europas Metropolregionen vor neuen Herausforderungen', *Informationen zur Raumentwicklung*, (11/12): 645–656, Bonn: BBR.

Göppel, K. (1993) 'Vernetzung und Kooperation – das neue Leitziel der Landesplanung', *Raumforschung und Raumordnung*, 52(2): 101–104.

Gorges, M. (2001) 'New institutionalist explanations for institutional change; a note of caution', *Politics*, 21(2): 137–145.

GOYH (Government Office for Yorkshire and the Humber) (1997) Yorkshire and the Humber Objective 2 Programme 1997–99 Single Programming Document.

—— (1998) Interview held at GOYH in June 1998.

Graham, D. and Hebbert, M. (1999) 'Greater London', in P. Roberts, K. Thomas and G. Williams (eds) *Metropolitan Planning in Britain*, London: Jessica Kingsley.

Gualini, E. (2000) 'Local development initiatives and regional development strategies: "new programming" and the influence of trans-national discourses in the reform of territorial policies in Italy', paper presented to the 3rd EURS conference, Voss, September 2000.

Hajer and Zonneveld (2000) 'Spatial planning in the network society', *European Planning Studies*, 8(3): 337–355.

Hall, P. (1994) 'Retrospect and prospect', in J. Simmie (ed.) *Planning London*, London: UCL Press.

—— (1998) 'Cities of Europe – motors in global economic competition', keynote speech at European Commission Transnational Seminar, *Cities: Perspective of the European System*, Lille.

Hampton, W. (1987) *Local Government and Urban Politics*, London: Longman.

Harding, A. (1994) 'Urban regimes and growth machines: towards a cross national research agenda', *Urban Affairs Quarterly*, 29: 295–317.

—— (2000) *The Devolution Agenda in England Overview for International Lessons for Devolution in England*. Conference, London, October.

——, Evans, R., Parkinson, M. and Garside, P. (1996) *Regional Government in Britain. An Economic Solution?*, Policy Press: University of Bristol.

Harris, N. (1997) 'Cities in a global economy: structural change and policy reactions', *Urban Studies*, 34(10): 1693–1703.

Harvey, D. (1995) 'Globalization in question', *Rethinking Marxism*, 8(4): 1–7.

—— (1996) *Justice, Nature and the Geography of Difference*, Oxford: Blackwell.

—— (2000) *Spaces of Hope*, Edinburgh: Edinburgh University Press.

Harvey, T. (1998) 'Portland, Oregon: regional city in a global economy', *Urban Geography*, 17(1): 95–114.

Hassemer, V. (1995) 'Landesentwicklungsplanung im Raum Berlin-Brandenburg. Ansprüche und Konzepte für die deutsche Hauptstadtregion aus Berliner Sicht', in W. Suess (ed.) *Hauptstadt Berlin*, vol. 2, Berlin: Berlin Verlag, pp. 349–363.

Hauswirth, I., Herrschel, T. and Newman, P. (2000) *Incentives and Disincentives to City-Regional Co-operation: The Case of Berlin-Brandenburg*. Regional Studies Association Conference, Aix-en-Provence, September.

Healey, P. (2000) 'New partnerships in planning and implementing future-oriented development in European metropolitan regions', *Informationen zur Raumentwicklung*, (11/12): 745–750, Bonn: BBR.

—— et al. (1997) *Making Strategic Spatial Plans*, London: UCL Press.

Hebbert, M. (1998) *London*, Chichester: Wiley.

Heeg, S. (1998) 'Vom Ende der Stadt als staatliche Veranstaltung. Reformulierung städtischer Politikformen am Beispiel Berlins', *PROKLA*, 28: 5–23.

—— (2001) *Politische Regulation des Raumes. Metropolen – Regionen – Nationalstaat*. Berlin: Edition Sigma.

Heide, H.-J. von der (1994) 'Stellung und Funktion der Kreise', in R. Roth and H. Wollmann (eds) *Kommunalpolitik – Politisches Handeln in den Gemeinden*, Opladen: Leske und Budrich, pp. 109–121.

Heidenreich, M. (1996) 'Beyond flexible specialization: the rearrangement of regional production orders', in Emilia-Romagna and Baden-Württemberg, *European Planning Studies*, 4(4): 401–419.

Heinze, R. and Voelzkow, H. (eds) (1997) *Regionalisierung der Strukturpolitik in NRW*, Opladen: Westdeutscher Verlag.

Held, D., McGrew, A., Goldblatt, D. and Perraton, J. (1999) *Global Transformations*, Cambridge: Polity Press.

Herrschel, T. (1997) 'Economic transformation, locality and policy in eastern Germany', *Applied Geography*, 17(4): 267–281.

—— (1998) 'From socialism to post-Fordism: the local state and economic policies in eastern Germany', in T. Hall and P. Hubbard (eds) *The Entrepreneurial City*, Chichester: Wiley, pp. 173–198.

—— (2000) 'Regions and regionalization in the five new *Länder* of eastern Germany', *European Urban and Regional Studies*, 7(1): 63–68.

—— and Newman, P. (2000) 'New regions in England and Germany: an examination of the interaction of constitutional structures, formal regions and informal institutions', *Urban Studies*, 37(7): 1185–1202.

Hirst, P. and Thompson, G. (1999) *Globilization in Question*, Cambridge: Polity Press.

HMSO (1999) *Greater London Authority Act*.

Hoffmann, A., Klatt, H. and Reuter, K. (1991) *Die Neuen Deutschen Bundeslander*, Bonn: Bonn Aktuell.

Holstila, E. (1999) 'The outlines of an urban policy for Finland', *Kvartti Quarterly*, 2(99): www.hel.fi/tietokeskus/en/kvartti/1999/2/kveero299.html, 'City of Helsinki Urban Facts' accessed on 12th April 2001.

Horváth, G. (1999) 'Changing Hungarian regional policy and accession to the European Union', *European Urban and Regional Studies*, 6(2): 166–177.

Hosse, O. and Schübel, S. (1996) 'Neue Ansätze der Regionalplanung und-politik in Thüringen', *Raumforschung und Raumordnung*, 54(4): 235–247.

House, J. W. (1981) 'Frontier studies: an applied approach', in A. Burnett and P. Taylor (eds) *Political Studies from Spatial Perspectives*, London: Wiley, pp. 291–312.

Hudson, R. (1999) 'The new economy of the new Europe: eradicating divisions or creating new forms of uneven development?', in R. Hudson and A. Williams (eds) *Divided Europe – Society and Territory*, London: Sage, pp. 29–63.

Huebner, M. (1995) '"Regionalisierung" von Unten: der Kommunalverbund Niedersachsen/Bremen', *Raumforschung und Raumordnung*, 53(3): 183–195.

IAW (Institut für angewandte Wirtschaftsforschung) (ed.) (1992) 'Die deutsch-polnischen Grenzgebiete als regionalpolitisches Problem', *IAW Forschungsreihe* 2/92, Berlin: IAW.

Imrie, R. and Thomas, H. (1999) *British Urban Policy*, 2nd edn, London: Sage.

Inforegio News (1999) European Commission Directorate-General for Regional Policies (EC DG XVI) (1999) Inforegio News, Newsletter no. 68.

IWH (Institut für Wirtschaftsforschung Halle) (ed.) (1997) *Strukturanalyse Sachsen Anhalt*, special volume, no. 2. Halle: IWH.

Jensen, O. and Richardson, T. (1999) *Constructing Spaces of Mobility and Polycentricity; The European Spatial Development Plan as a New Power/Rationality Nexus for European Spatial Planning*, Aesop Congress, Bergen.

Jensen-Butler, C., Shachar, A. and van Weesep, J. (eds) (1997) *European Cities in Competition*, Avebury: Aldershot.

Jessop, B. (1990) 'Regulation theories in retrospect and prospect', *Economy and Society*, 19(2): 153–216.

—— (1994) 'Post-Fordism and the state', in A. Amin (ed.) *Post-Fordism: A Reader*, Oxford: Blackwell, pp. 251–279.

—— (1995) 'The regulation approach, governance and post Fordism', *Economy and Society*, 24, 307–333.

—— (1997) 'Capitalism and its future: remarks on regulation, government and governance', *Review of International Political Economy*, 4(3): 561–581.

—— (1998) 'The rise of governance and the risks of failure: the case of economic development', *International Social Science Journal*, (155): 29–45.

John, P. (2000) 'The Europeanisation of sub-national governance', *Urban Studies*, 37(5/6): 877–894.

—— (2001) *Local Governance in Western Europe*, London: Sage.

—— and Cole, A. (2000) 'When do institutions, policy sectors and cities matter? Comparing networks of local policy-makers in Britain and France', *Comparative Political Studies*, December.

Jonas, A. and Wilson, D. (1999) *The Urban Growth Machine: Critical Perspectives Two Decades Later*, Albany: State University of New York Press.

Jones, M. (1998) *New Institutional Spaces*, London: Jessica Kingsley.

—— and MacLeod, G. (1999) 'Towards a regional renaissance? Reconfiguring and rescaling England's economic governance', *Transactions*, 24(3): 295–314.

Jouve, B. and Lefèvre, C. (eds) (1999) *Villes, Métropoles. Les nouveaux territoires du politique*, Paris: Anthropos.

Karrenberg, H. (1991) 'Die Finanzierung der Kommunalhaushalte in den Neuen Ländern', in HWWA Institut für Wirtschaftforschung (ed) *Zeitschrift für Wirtschaftspolitik*. Wirtschaftsdienst VI, Hamburg, pp. 296–303.

Kearns, A. and Paddison, R. (2000) 'New challenges for urban governance: introduction to the review issue', *Urban Studies*, 37(5/6): 845–850.

Keating, M. (1997) 'The invention of regions: political restructuring and territorial government in western Europe', *Environment and Planning C*, 15: 383–398.

Keil, R. (1998) *Los Angeles: Globalization, Urbanization and Social Struggles*, Chichester: Wiley.

Kennedy, R. (1991) *London World City*, London: HMSO.

Kerstens, M. (1998) 'The struggle of Dutch transport regions: life and death of a regional planning concept', *European Planning Studies*, 6(3): 299–313.

Kistenmacher, H., Geyer, Th. and Hartmann, P. (1994) *Regionalisierung in der kommunalen Wirtschaftsförderung*, Cologne: Deutscher Gemeindeverlag/Kohlhammer.

Koch, A. and Fuchs, G. (1999) 'Economic globalization and regional penetration: the failure of networks in Baden-Württemberg', *European Journal of Political Research*, 31(1): 57–75.

Krätke, S. (1996) 'Where East meets West. The German–Polish border region in transformation', *European Planning Studies*, 4(6): 647–669.

—— (1998) 'Problems of cross-border regional integration: the case of the German–Polish border area', *European Urban and Regional Studies*, 5(3): 249–262.

—— (1999) 'The German–Polish border region in a new Europe', *Regional Studies*, 33(7): 631–642.

Krumbein, W. (1997) 'Industrie- und Regionalisierungspolitik in Niedersachsen. Koordinierungsprobleme bei der Konzeptionierung und Umsetzung von Reformvorhaben', in U. Bullmann (ed.) *Die Politik der Dritten Ebene*, Baden-Baden: Nomos, pp. 364–377.

Kunzmann, K. (1996) 'Euro-megalopolis or themepark Europe', *International Planning Studies*, 1(2): 143–163.

Lambooy, J. (1998) 'Polynucleation and economic development: the Randstad', in Land Saxony-Anhalt (1998) 'Gesetz- und Verordnungsblatt für das Land Sachsen-Anhalt, 4 May 1998, no. 16, pp. 237–262, Publication of *Land* Planning Act (Landesplanungsgesetz des Landes Sachsen-Anhalt' LPlG) of 28 April 1998).

Landesamt für Statistik, Brandenburg (1999) 'Bevölkerungsentwicklung im engeren Verflechtungsraum Brandenburg-Berlin von 1991–1998'. Press Note 85/99, 1 July 1999.

Lauria, M. (ed.) (1997) *Reconstructing Urban Regime Theory*, Thousand Oaks, CA: Sage.

LDA (London Development Agency) (2000) *Draft Economic Development Strategy*, London: LDA.

LDP (London Development Partnership) (2000) 'Building London's economy', London: LDP, unpublished.

LEDA (Leeds Development Agency) (1996) 'Leeds economic development strategy', Leeds: unpublished.

—— (1998) *Economic Development Strategy*, Leeds: LEDA.

Lefèvre, C. (1998) 'Metropolitan government and governance in western countries: a critical review', *IJURR*, 22(1): 9–25.

Le Galès, P. (1998) 'Regulations and governance in European cities', *International Journal of Urban and Regional Research*, 22: 482–506.

—— (2000) 'Private sector interests and urban governance', in A. Bagnasco and P. Le Galès (eds) *Cities in Contemporary Europe*, Cambridge University Press.

—— and John, P. (1997) 'Is the grass greener on the other side? What went wrong with French regions and the Implications for England', *Policy and Politics*, 25: 51–60.

—— and Lequesne, C. (eds) (1998) *Regions in Europe*, London: Routledge.

Lenhardt, K. (1998) 'Bubble politics in Berlin', *PROKLA*, 28: 41–66.

Lever, W. F. (1997) 'Delinking urban economies: the European experience', *Journal of Urban Affairs*, 19(2): 227–238.

Lieberda, E. (1996) *Regionalentwicklung in Grenzregionen: Eine Euroregio als regionale Entwicklungsstrategie?*, Passau: Passavia Universitätsverlag.

Lipietz, A. (1992) 'A regulationist approach to the future of urban ecology', *Capitalism, Nature, Socialism*, 3(3): 101–110.

—— (1995) 'Avoiding megapolization. The battle of Île-de-France', *European Planning Studies*, 3(2): 143–154.

Llewelyn-Davies (1996) *Four World Cities: A Comparative Study of London, Paris, New York and Tokyo*, London: Government Office for London.

Logan, J. and Molotch, H. (1987) *Urban Fortunes*, Berkeley: University of California Press.

——, Whaley, R. and Crowder, K. (1997) 'The character and consequences of growth regions: an assessment of twenty years of research, Paper to Milan Construction Conference, University of Turku.

Lovering, J. (1999) 'Theory led by policy: the inadequacies of the "new region-alism"', *International Journal of Urban and Regional Research*, 23: 379–395.

LPAC (2000) *Co-ordinating Regional Spatial Planning Between London, the South East and East Regions: Advice to the Mayor*, Committee Report 23/2000.

Lutzky, N., Lorentz, D., Morrish, S. and Dütz, A. (1999) European metropolitan regions project. Strategies for sustainable development of European metro-politan regions. Evaluation report. Submitted to the European Regional Confer-ence 'European metropolitan regions', Essen, September 1999, unpublished.

MacLeod, G. (1999) 'Place, politics and place dependence', *European Urban and Regional Studies*, 6(3): 231–253.

—— (2000) 'The learning region in an age of austerity: capitalizing on knowledge, entrepreneurialism and reflexive capitalism', *Geoforum*, 31: 219–236.

Marinetto, M. (2001) 'The settlement and process of devolution: territorial politics and governance under the Welsh Assembly', *Political Studies*, 49(2): 306–322.

Marks, G. (1993) 'Structural policy and multilevel governance in the EC', in A. Cafruny and G. Rosenthal (eds) *The State of the European Community Vol. 2*, Harlow: Longman.

Marshall, T. (1996) 'Barcelona – fast forward? City entrepreneurialism in the 1980s and 1990s', *European Planning Studies*, 4(1): 147–165.

—— (2000) 'Urban planning and governance – is there a Barcelona model?', paper presented to the RGS-IBG Annual Conference, Brighton, January 2000.

Martin, R. (1999) *The Regional Dimension in European Public Policy. Convergence or Divergence?*, Basingstoke: Macmillan.

Martinelli, F. (1998) 'The governance of development and policy in southern Italy. Notes for a critical reappraisal', paper presented to the 2nd EURS Conference, Durham, 1998.

Massey, D. (1978) 'Regionalism: some current issues', *Capital and Class*, (6): 106–125.

Mawson, J. (1997a) 'The English regional debate. Towards regional governance or government?', in J. Bradbury and J. Mawson (eds) *British Regionalism and Devolution*, London: Jessica Kingsley.

—— (1997b) 'New Labour and the English regions. A missed opportunity?', *Local Economy*, November, pp. 194–203.

McNeill, D. (2001) 'Embodying a Europe of the Cities: Geographies of Mayoral Leadership', *Area*, 33(4): 353–359.

Miosga, M. (1999) 'Europäische Regionalpolitik in Grenzregionen', *Münchner Geographische Hefte*, vol. 79, Passau: L.I.S.

Molotch, H. (1976) 'The city as a growth machine: toward a political economy of place', *American Journal of Sociology*, 82: 309–332.

Morata, F. (1997) 'The Euro-region and the C6 network: the new politics of subnational cooperation in the West-Mediterranean area', in M. Keating and J. Loughlin (eds) *The Political Economy of Regionalism*, London: Frank Cass, pp. 292–305.

MRLU (Ministerium für Raumordnung, Landwirtschaft und Umwelt) Saxony-Anhalt (ed.) (1996) 'Landesentwicklungsbericht 1996', Magdeburg: unpublished.

MRU (Ministerium für Raumordnung und Umwelt) Saxony-Anhalt (1999) 'Landes-entwicklungsplan für das Land Sachsen Anhalt 1999', Magdeburg: unpublished.

Müller, B. (1995) 'Impulse aus dem Osten? Erfahrungen und Perspektiven der Regionalplanung in den ostdeutschen Ländern', *ARL* (*Akademie für Raum-forschung und Landesplanung): Zukunftsaufgabe Regionalplanung.* Wissenschaft-liche Plenarsitzung (Academic Plenary), Chemnitz, Hanover: ARL, pp. 31–52.

—— (1999) 'Kooperative Entwicklungsansätze in Ostdeutschland. Von der Raum-ordnung zur Regionalentwicklung', *Informationen zur Raumentwicklung*, 9/10: 597–615.

MUNR/SENSUT (Ministerium für Umwelt, Naturschutz und Raumordnung des Landes Brandenburg/ Senatsverwaltung für Stadtentwicklung, Umweltschutz und Technologie) (1998) *Gemeinsam Planen für Berlin und Brandenburg*, Berlin and Potsdam: MUNR/SENSUT.

Négrier, E. (2000) 'French regulatory path? state, economy and territory', *TESG* (*Tidjschrift voor Economische en Sociale Geografie*), 91(3): 248–262.

Nevin, E. (1990) *The Economics of Europe*, Basingstoke: Macmillan.

Newman, P. (1995) 'London pride', *Local Economy*, 10: 117–123.

—— and Thornley, A. (1996) *Urban Planning in Europe. International Competition, National Systems and Planning Projects*, London: Routledge.

—— (1997) 'Fragmentation and centralization: influencing the urban policy agenda in London', *Urban Studies*, 34: 967–988.

OECD (1999) *Draft Priciples of Metropolitan Governance*, Paris: OECD.

Ohmae, K.-I. (1995) *The End of the Nation State: The Rise of Regional Economies*, New York: Harper Collins.

Orfield, M. (1997) *Metropolitics. A Regional Agenda for Community and Stability*, Washington: Brookings Institution.

Painter, J. (1991) 'Regulation theory and local government', *Local Government Studies*, 17(6): 23–44.

—— (1995) 'Regulation theory, post-Fordism and urban politics', in D. Judge, G. Stoker and H. Wollmann (eds) *Theories of Urban Politics*, London: Sage.

Parkinson, M., Bianchini, F., Dawson, F., Evans, R. and Harding, A. (1992) *Urbanisa-tion and the Functions of Cities in the European Community*, Brussels: EC.

Pastor, M., Dreier, P., Grigsby, J. and Lopez-Garza, M. (2000) *Regions that Work. How Cities and Suburbs Can Grow Together*, Minneapolis: University of Wisconsin Press.

Peck, J. (1998) 'Geographies of governance: TECs and the neo-liberalism of "local interests"', *Space and Polity*, 2: 5–31.

Perkmann, M. (1999) 'Building governance institutions across European borders', *Regional Studies*, 33(7): 657–668.

Petzold, S. (1994) 'Zur Entwicklung und Funktion der kommunalen Selbstver-waltung in den neuen Bundesländern', in R. Roth and H. Wollmann (eds) *Kommunalpolitik – Politisches Handeln in den Gemeinden*, Opladen: Leske und Budrich, pp. 34–51.

Pierce, N. R. (1993) *Citistates. How Urban America Can Prosper in a Competitive World*, Washington, DC: Seven Locks Press.

Pierson, P. (2000) 'Increasing returns, path dependence, and the study of politics', *American Political Science Review*, 94(2): 251–266.

Priebs, A. (1999) 'Die Region ist die Stadt! Ein Plädoyer für dauerhafte und verbindliche Organisationsstrukturen für die Stadtregion', *Informationen zur Raumentwicklung*, 9/10, Bonn: BBR.

Priemus, H. (1998) 'The Randstad and the central Netherlands urban ring: planners waver between two concepts', *European Planning Studies*, 6(4): 443–455.

Regt, A. de and Burg, A. van der (2000) 'Randstad Holland: the deltametropolis', *Informationen zur Raumentwicklung*, (11/12): 691–704, Bonn: BBR.

Robert, J. (1997) 'Raumordnerische Zusammenarbeit in Europa aus der Sicht Frankreichs am Beispiele des Nordwesteuropäischen Metropolraums', *Informationen zur Raumentwicklung*, (6/97): 419–422.

Rodriguez-Pose, A. (1996) 'Growth and institutional change: the influence of the Spanish regionalisation process on economic performance', *Environment and Planning C: Government and Policy*, 14: 71–87.

—— (1998) *The Dynamics of Regional Growth in Europe. Social and Political Factors*, Oxford: Clarendon Press.

Ross, B. and Levine, M. (2000) *Urban Politics: Power in Metropolitan America*, Itasca, IL: F. E. Peacock.

Sallez, A. (1998) 'France', in L. van den Berg and J. van der Meer (eds) *National Urban Policies in the European Union*, Ashgate: Aldershot, pp. 97–131.

Savitch, H. and Vogel, R. (1996) *Regional Politics*, Thousand Oaks, CA: Sage.

SCC (1998) Interview with Sheffield City Council, September 1998.

Schädlich, M. (1997) 'Regionalkonferenz Halle-Leipzig. Ergebnisse, Erfahrungen, Perspektiven', *Informationen zur Raumentwicklung*, (3): 167–176.

Schmidt-Eichstaedt, G. (1994) 'Die Kommunen zwischen Autonomie und (Über-) Regelung durch Bundes- und Landesrecht sowie durch die EG-Normen', in R. Roth and H. Wollmann (eds) *Kommunalpolitik – Politisches Handeln in den Gemeinden*, pp. 95–108.

Schmitz, G. (1995) 'Regionalplanung', in *Handwörterbuch der Raumordung*, Hanover: ARL, pp. 823–830.

Scholich, K. (1995) 'Regionale Entwicklungskonzepte und Regionalkonferenzen: Instrumente einer regionalisierten Landesentwicklungs- und Regionalpolitik', in *ARL (Akademie für Raumordnung und Landesplanung): Zukunftsaufgabe Regionalplanung*. Anforderungen, Analysen, Empfehlungen, Hanover: ARL.

Schulte, W. (2000) 'Die gemeinsame Landesplanung für den Metropolenraum Berlin-Brandenburg. Institutionelle Absicherung einer integrierten Siedlungsstruktur- und Verkehrsplanung für eine nachhaltige Entwicklung', *Informationen zur Raumentwicklung*, (11/12): 705–712, Bonn: BBR.

Scott, A. J. (1988) *New Industrial Spaces*, London: Pion.

—— (1998) *Regions and the World Economy*, Oxford: Oxford University Press.

Scott, J., Sweedler, A., Ganster, P. and Eberwein, W. (eds) (1996) *Border Regions in Functional Transition: Europeans and North American Perspectives*, Berlin: Institut für Regionalentwicklung und Strukturentwicklung Erkner.

Scott, W. (1997) 'Promoting transboundary regionalism on the German Polish border: some preliminary observations', paper presented to the Regional Studies Association Conference, Frankfurt/Oder (Germany), September 1997.

SEEDA (2000) *Building a World Class Region: An Economic Strategy for the South East of England*, Guildford: Seeda.

Seixas, J. (1996) 'The future of governance in Lisbon city region: expected urban politics for the XXIst century', mimeograph.

SENSUT (Senatsverwaltung für Stadtentwicklung, Umweltschultz und Technologie) (1996) 'Stadt und Nachbarn. Kommunale und regionale Zusammenarbeit im Spree-Havel-Raum', Berlin: SENSUT, unpublished.

Shutt, J. (2001) 'New regional development agencies in England: wicked issues', in J. Gibney and G. Bentley (eds) *Building a Competitive Region. RDAs and Business Change*, Aldershot: Ashgate.

—— and Colwell, A. (1997) *Towards 2006*, London: Local Government Information Unit.

Simmons, M. (2000) 'New London government and its spatial development strategy', *Informationen zur Raumentwicklung*, (11/12): 671–678, Bonn: BBR.

Soja, E. W. (2000) *Postmetropolis: Critical Studies of Cities and Regions*, Oxford: Blackwell.

Sotarauta, M. and Linnamaa, R. (1998) 'Finnish multi-level policy-making and the quality of local development policy processes: the cases of Oulu and Seinänaapurit sub-regions', *European Planning Studies*, 6(5): 505–523.

SPESP (2000) 'Study programme on European spatial planning, final report', BBR *Forschungen*, no. 103.2.

Stiens, G. (2000) 'Regionale Regulation und faktische Auflösung überregionaler Raumordnung? Die "deutschen Europäischen Metropolregionen" als Fall', *Informationen zur Raumentwicklung*, (9/10): 517–536. Bonn: BBR.

Stone, C., Orr, M. and Imbroscio, D. (1991) 'The reshaping of urban leadership in US cities: a regime analysis', in M. Gottdiener and C. Pickvance (eds) *Urban Life in Transition*, Newbury Park, CA: Sage.

Storper, M. (1997) *The Regional World: Territorial Development in a Global Economy*, London: Guildford Press.

Strom, E. (1996) 'The political context of real estate development: central city rebuilding in Berlin?', *European Urban and Regional Studies*, 3(1): 3–17.

—— (2001) *Building the New Berlin*, Oxford: Lexington Books.

Sturm, P. (2000) 'Region Frankfurt/Rhein Main', *Informationen zur Raumentwicklung*, (11/12): 6705–6712, Bonn: BBR.

Swanstrom, T. (1996) 'Ideas matter: reflections on the new regionalism', *Cityscape*, 2: 5–21.

Swyngedouw, E. (1996) 'Reconstructing citizenship, the re-scaling of the state and the new authoritarianism: closing the Belgian mines', *Urban Studies*, 33(8): 1499–1521.

—— (1997) 'Neither global nor local: "glocalization" and the politics of scale', in K. R. Cox (ed.) *Spaces of Globalization*, London: Guildford Press, pp. 137–166.

—— (2000) 'Authoritarian governance, power and the politics of rescaling', *Environment and Planning D: Society and Space*, 18: 63–76.

Syrett, S. and Silva, M. (1999) 'New institutions for new forms of governance: regional development agencies in Portugal', paper presented to the International Conference of the Regional Studies Association, Bilbao, September 1999.

—— (2001) 'Regional development agencies in Portugal', *Regional Studies*, 35(2): 174–180.

Taylor, P. J. (1997) 'Is the United Kingdom big enough for both London and England?', *Environment and Planning*, 29(5): 766–770.

—— and Hoyler, M. (2000) 'The spatial order of European cities under conditions of contemporary globalisation', *Tijdschrift voor Economische en Sociale Geografie*, 91(2): 176–189.

Thornley, A. (2001) *The New Government of London: Reflections on its Origins and Implications*, AESOP, ACSP conference, Shanghai.

TLG (Thuringa *Land* Government) (1998) Ministry of Planning, interview held June 1998.

Tomaney, J. and Ward, N. (2000) 'England and the "New Regionalism"', *Regional Studies*, 34: 417–478.

Travers, T. and Jones, G. (1997) *The New Government of London*, York: Joseph Rowntree Foundation.

Tsoulouvis, L. (1999) 'Rethinking the relationship between regional policy in the new Europe and national/regional systems of spatial planning', paper to Regional Studies Association Conference, Bilbao.

Turok, I. and Bachtler, J. (eds) (1997) *The Coherence of EU Regional Policy*, London: Jessica Kingsley.

Uusimaa Regional Council (2001) Information from their website: www.uuden-maanliitto.fi/eng/urc_keski.htm, accessed on 4 December 2001.

Veer, J. van der (1998) 'Metropolitan government in Amsterdam and Eindhoven: a tale of two cities', *Environment and Planning C: Government and Policy*, 16: 25–50.

Veltz, P. (2000) 'European cities in the world economy', in A. Bagnasco and P. Le Galès (eds) *Cities in Contempory Europe*, Cambridge: Cambridge University Press, pp. 33–47.

Voelzkow, H. (2000) 'Regieren im Europa der Regionen. Vom Wohlfahrtsstaat zum Wettbewerbsstaat, vom Makro-korporatismus zum Meso-korporatismus', *Informationen zur Raumentwicklung*, (9/10): 507–516, Bonn: BBR.

Ward, K. G. (1997) 'Coalitions in urban regeneration: a regime approach', *Environment and Planning A*, 29: 1493–1506.

Warleigh, A. (1997) 'A committee of no importance? Assessing the relevance of the Committee of the Regions', *Politics*, 17(2): 101–108.

Weichhart, P. (2000) 'Designerregionen – Antworten auf die Herausforderungen des Globalen Standortwettbewerbs', *Informationen zur Raumentwicklung*, (9/10): 549–566, Bonn: BBR.

Wild, S. (1997) 'The Northern League: the self representation of industrial district in their search for regional power', *Politics*, 17(2): 95–100.

Williams, R. (1996) *European Union Spatial Policy and Planning*, London: Paul Chapman.

Wise, M. (2000) 'From Atlantic Arc to Atlantic Area: a case of subsidiarity against the regions', *Regional Studies*, 34(9): 865–873.

Wollmann, H. and Lund, S. (1998) 'European integration and the local authorities in Germany: impacts and perceptions', in M. Goldsmith and K. Klausen (eds) *European Integration and Local Government*, Cheltenham: Edward Elgar, pp. 57–74.

Y&H RDA (Yorkshire and the Humber Regional Development Agency) (1999) Interview held with officer on 7 July 1999.

Zarth, M. (1997) 'Was macht Regionalkonferenzen erfolgreich?', *Informationen zur Raumentwicklung*, 3: 155–160.

Zonneveld, W. and Faludi, A. (eds) (1997) 'Vanishing borders: the second Benelux structural outline', *Built Environment*, Special edition, 23(1): 5–81.

Index